Date Due

APR 2 3 1990			

SOUND UNDERWATER

SOUND UNDERWATER

Gregory Haines

DAVID & CHARLES : NEWTON ABBOT LONDON

CRANE RUSSAK : NEW YORK

First published 1974 by
David & Charles (Holdings) Limited
South Devon House Newton Abbot Devon
ISBN 0 7153 6219 4

Published in the United States of America by
Crane Russak & Company, Inc.
52 Vanderbilt Avenue New York New York 10017
ISBN 0 8448 0302 2
Library of Congress Catalog Card
Number 73 92386

Set in 11 on 13 point Times
and printed in Great Britain
by Latimer Trend & Company Ltd
for David & Charles (Holdings) Limited
South Devon House Newton Abbot Devon

CONTENTS

LIST OF ILLUSTRATIONS

PLATES

LIST OF ILLUSTRATIONS

FIGURES

SOUND UNDERWATER

1

BASIC CONCEPTS

THE SEA

THE EARTH IS UNIQUE AMONG THE PLANETS OF THE SOLAR SYSTEM in being largely covered by water. All the oceans and the seas together occupy a volume of 324 million cu miles and cover 71 per cent of the earth's surface. There is enough water, in fact, to fill a cylinder about 2 miles in diameter stretching all the way from here to the sun. This is the size of the test tube in which, during the last few million years, all forms of life have evolved.

Today, necessity is rapidly replacing academic curiosity as the motivation for undersea exploration. The population explosion is making ever-increasing demands on the sources of protein available in the sea. The efficient exploitation of all species of edible fish, as well as research into the developing science of fish farming, requires a knowledge of the sea. The extraction of minerals from the sea and seabed and offshore drilling for oil and gas involve working in the sea and understanding it.

In all branches of oceanography it is necessary to make measurements and observations underwater, to draw maps of the seabed and of what lies beneath it, and to monitor and control instruments and installations in mid-water and on the bottom. The sea, by its nature, imposes peculiar difficulties for all these activities. For the diver it is the pressure, which increases by 1 atmosphere for every 10m (33ft) of depth, that will set the ceiling, or perhaps it would be more appropriate to say the floor, on his depth of penetration. A still greater handicap for all forms of underwater investigation results from the rapid attenuation of electro-magnetic waves in water. Light also has very limited penetration. Even the exceptionally clear waters of the Bahamas are roughly equivalent in transparency to a moderately heavy fog on land.

Electric and magnetic fields and electro-chemical ionisation effects can be used for obtaining information over short distances but lack adequate penetration or resolution. The only other phenomenon that can be harnessed in the sea as a substitute for light and radio transmission is sound, with a rate of absorption that is less, and a speed roughly four times as great, as in air.

UNDERWATER SOUND

The earliest reference to sound underwater comes from the notebooks of Leonardo da Vinci, who wrote: 'If you cause your ship to stop, and place the head of a long tube in the water, and place the other extremity to your ear, you will hear ships at a great distance from you.' As a hydrophone, such a device is obviously inefficient, owing to the mis-match between the water and the tube, and there would be no indication of either direction or range.[1] Even so, this crude arrangement is better than nothing, since the interface between sea and air acts as a virtually complete barrier to sound. Some idea of direction can be obtained by adding a second tube, bringing the upper end of this to the other ear and then rotating the assembly until the intensity of sound in each ear is matched. This binaural technique was used as recently as World War I for submarine detection.

The next event usually quoted in the history of underwater acoustics is the first attempt to measure the speed of sound in water. This was done in 1826 by Colladon and Sturm in the Lake of Geneva by striking an underwater bell and simultaneously emitting a light flash. The time interval between the flash and the sound signal was measured at a known distance and the speed of 1435m/sec was calculated. This is a slight underestimate but reasonably accurate in the circumstances. Submerged bells were being used at the turn of the century for navigation; suspended below lightships and buoys they could be detected in poor visibility by a microphone fitted in the ship's hull.

Two important discoveries, both fundamental to underwater acoustic technology, date from the 1800s. James Joule in 1840 noticed that certain substances when magnetised have the property of changing length and, conversely, that changes in the magnetic properties can be induced by mechanical stress; and he measured this magnetostrictive effect, which is found in iron, nickel, cobalt, manganese and their alloys. In 1880 Jacques and Pierre Curie discovered that certain crystalline substances develop an electric charge when under stress, and this also works both ways, ie, a voltage potential across the face of the material causes a small change in length in the same plane. This is known as the piezoelectric effect. Quartz crystals, Rochelle salt and ammonium dihydrogen phosphate (ADP) are the best known among a number of substances exhibiting it. Both these phenomena provide means of initiating sound waves underwater at any desired frequency, although the techniques for doing so were not developed until many years after they were discovered.

Sound waves are pressure waves. The particles of the material through which the sound is travelling oscillate longitudinally, ie, in the direction of propagation. The speed of sound depends on the density and elasticity of the material.[1] The exact speed in water varies slightly with temperature, pressure and salinity.[2]

Considering a point source of sound in a homogeneous medium, the sound waves emanating from it in all directions can be visualised as a series of expanding concentric spheres, the surfaces of which represent at any instant of time regions of maximum compression. Midway between these regions there are zones in which the molecules are fully stretched. In both these areas sound is in the form of the potential energy of the strains set up in the elastic medium. In changing from a state of compression to one of tension and back again a state of maximum kinetic energy is traversed. Thus the energy contained in the sound wave is constantly changing between potential and kinetic energy as it moves forward. For most purposes it is convenient to consider this energy transference process in terms of a sine wave (Fig 1). In this diagram the energy of the wave is proportional to the square of the maximum acoustic pressure or amplitude (A). The intensity (I) is defined as the energy crossing a unit area at right-angles to the direction of propagation, per unit time, measured in Hz.[3] The speed (c) is related to frequency and wavelength (λ—the Greek letter l, pronounced 'lambda') by the expression $c = \lambda f$.

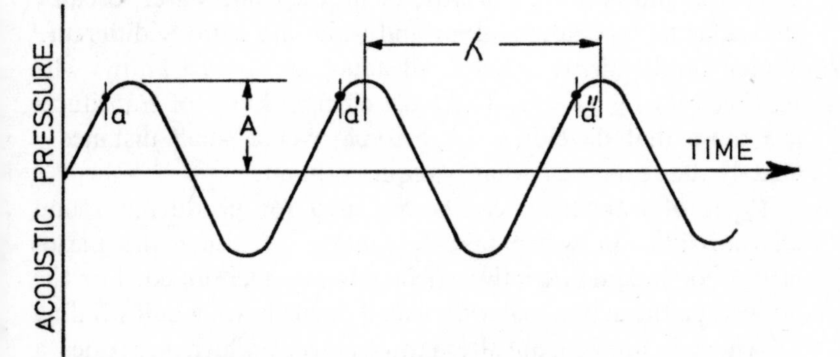

FIG 1 A sound wave is shown graphically in the form of a sine curve. The energy in the wave is proportional to the square of the acoustic pressure, measured vertically, or amplitude (A). The points a, a′, a″ occur in the same relative positions on the wave, and are said to be in phase

One other important concept is the phase angle, which is defined by the point on the curve reached by the wave at any particular moment. In Fig 1 the points a, a′ and a″ have equal phase angles and are said to be in phase. When two different signals are in phase, their amplitudes add up, or reinforce each other; when out of phase they are subtracted; and in between these two limits the combined amplitude is the vector sum. The intensity of sound is often referred to in relative rather than absolute terms. The decibel (dB) is used to express the ratio between intensities.[4]

Sound waves lying within the audible frequency range are known as sonic and those above it as ultrasonic. The upper limit of normal hearing is usually said to be about 20kHz and the useful range of ultrasonic frequencies extends to about 500kHz for underwater use.

METHODS OF TRANSMISSION

Loudspeakers and microphones used for transmitting and receiving sound in air are of little or no use underwater, because the acoustic impedances of air and water are entirely different.[1] Water, unlike air, is a 'hard' substance, as anyone knows who has been hit by a wave. The most efficient kinds of transducer are those that develop a large force over a small distance—exactly the reverse of what is required in air.[5]

Types of transducer commonly used for producing sound transmissions in water take advantage of either the piezo-electric or magnetostrictive effects already mentioned. For the latter type the active material, usually nickel, has a coil winding to which is applied an alternating current which generates a changing magnetic field, causing the metal to expand and contract, ie, to oscillate at the frequency of the supply. The effect is independent of the direction of flow of the current, and so it is necessary to magnetise the element permanently to ensure that the oscillating magnetic field does not reverse the direction of magnetisation. If this were to happen, the frequency would be doubled in the process of converting from electrical to acoustic energy, and there would also be a loss of efficiency. Permanent magnets are normally used for this purpose, the active element being made up of a stack of thin stampings that must first be annealed to bring the metal to the correct state of pliability. It

happens that this process causes oxidisation, which, most conveniently, insulates the stampings from each other, thus avoiding eddy current losses.

In earlier types of magnetostriction transducers the nickel stampings were in the form of rings, toroidally wound. The whole assembly was mounted horizontally at the centre of a conical reflector. The sound wave was thus deflected into a generally vertical direction, this type of arrangement being used for echosounding. In more modern types the oscillating face of the transducer is flat and rectangular, and in direct contact with the sea. Piezoelectric transducers are also normally flat faced. A relatively recent development in transducer design is the use of certain polycrystalline ceramics such as barium titanate and lead zirconate titanate, which exhibit an electrostrictive effect very similar to those already described. These compounds have to be permanently polarised by subjecting them to a high electrostatic field.

To obtain the maximum response, transducers of these types are designed to oscillate at their resonant frequencies, this being determined by the dimensions of the active element in the direction of the oscillations. It is also possible to vary the sharpness of the resonance, or 'Q' factor, by changes in design.

The piezoelectric and electrostrictive types are more efficient, the energy conversion factor being of the order of 80 per cent as opposed to something under 50 per cent for magnetostrictive types, but the latter are cheaper to manufacture. There are of course many other complex factors to take into account in selecting the design best suited to any particular need. Transducer design has in recent years become a highly specialised branch of acoustic engineering, and is still in the stage of active development (see plate, p 34).

All these transducers are activated by a source of electrical energy, which is normally supplied from a power amplifier giving full control over the pulse length, frequency and power. For certain applications, where a burst of high power contained within a very short pulse length is required, a condenser discharge is still sometimes used. Pulses as short as a single wavelength can be obtained in this way.

For other uses special types of sonic projector are required. Explosive sources are employed for air/sea rescue position-fixing within the SOFAR (Sound Fixing and Ranging) channel, for echoranging on submarines with sono-buoys and for seismic investigations of sub-bottom structures. The acoustic energy is derived from the shock wave, and this is followed by a series of 'bubble pulses' caused by the oscillations of the gas envelope remaining after the explosion. Most of the sound energy is in the 10 to 100Hz frequency band; it is 15dB down at 1kHz. After travelling for long distances, as in the SOFAR application, the pressure signature is complicated by reflections from the surface and differential velocities. Types of explosive commonly used are TNT and an explosive mixture of gases, such as oxygen and propane. Other methods are high energy electrical discharge (sparkers), hydromechanical resonators and the sudden release of compressed air. Finally, there is a device known as a 'thumper', which consists of a helical coil embedded in the surface of an epoxy plate with a spring-loaded aluminium disk held immediately adjacent to the coil. A pulse of current in the coil induces eddy currents in the disk, causing it to be violently repelled by the coil, with consequent shock generation. Further details of these methods will be found in Chapter 5.

DIRECTIVITY

When transmitting sound pulses in order to obtain echoes from the seabed or from objects in mid-water, it is desirable to make the beam as directional as possible. The sound energy is then mainly concentrated into a narrow sector instead of being dissipated in all directions, and thus the intensity of the beam, when it reaches the target, is the greater, as is the returning signal. Provided that the angle subtended by the beam is not less than that subtended by the target, the ratio of echo strength to background noise will also be improved, giving the expectation of a greater maximum range of detection. A narrow beam also has the advantage of locating the target more accurately in bearing. For any given wavelength, the larger the face of the transducer, the narrower the transmitted beam becomes. This can be understood by considering the vibrating face as being made up of a series of separate point sources of sound from each of which a spherical wave train is emanating. In the direction at right-angles to the face all these sound waves will remain in phase and therefore reinforce each other, whereas in other directions they will be to a greater or lesser extent out of phase. Near the transducer, in what is called the 'near field' or Fresnel zone, the interference patterns created by all the adjacent waves are complex, but in the far field a pattern of intensity in relation to direction becomes established.[6] This is the typical beam pattern with a maximum power point at the centre, falling away on either side and then rising again to secondary peaks or 'side lobes' (Fig 2). These interference patterns, which lead to the formation of a directional beam, are typical of any kind of wave and depend only on the size of the transducer face in terms of the wavelength. In this respect there is no difference between sound waves and electromagnetic waves. It is beside the point that their respective speeds of propagation are vastly different (3.10^8 as against $1{\cdot}5.10^3$m/sec). For example, 'X' band radar and 50kHz sound waves both have a wavelength of 3cm, so if the dimensions of the aerial and transducer arrays are the same in each case, the two beam widths will also be the same. The width of the main lobe varies inversely as the size of the radiating surface measured in wavelengths.

It is a common fallacy to assume that the polar diagram (Fig 2) defines the shape of the beam. It simply shows how much energy is radiated on either side of the centre bearing by means of a graph in which energy is plotted vertically and bearing horizontally. Furthermore, the diagram only shows the energy

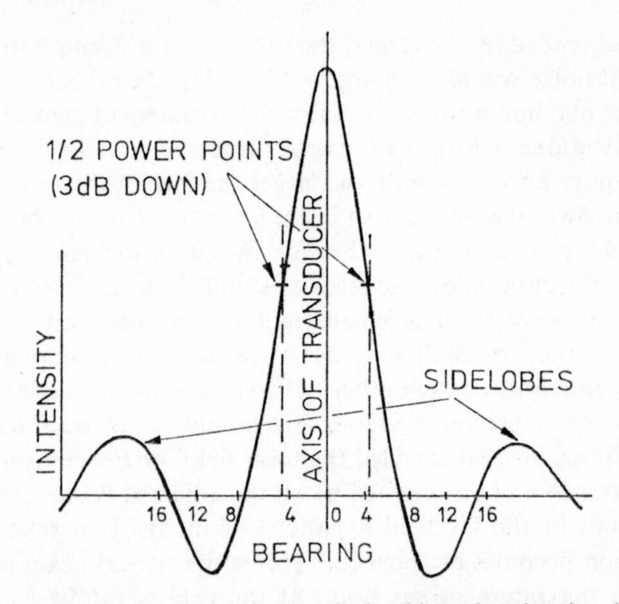

FIG 2 The graph of intensity (vertical axis) against bearing (horizontal axis), known as a polar diagram, indicates the degree of sharpness of a directional sound beam. The beam width is usually defined by the total angle between half-power points, ie, the sector over which the intensity is over half the peak value at right-angles to the transducer face

distribution in one particular plane. If the transducer face is circular, the diagram will be the same for all diameters. If it is elliptical, it will show the least concentrated beam across the minor axis, and the sharpest across the major axis. Similar results are obtained from a rectangular face except that across the diagonal the energy transmitted in the secondary lobes is relatively less. The reason for this will be found in the explanation of tapering, which follows. To obtain a mental image of the three-dimensional beam pattern one must think of the polar diagram on all bearings together, always remembering that the definition of the beam width is no more than a conventional way of describing it and does not imply any kind of limit or envelope. Echoes can be obtained from bearings outside the 'beam', always providing that the targets are near enough and have sufficient reflectivity.

In thinking of the impact of the returning echo on the transducer array similar considerations apply. If it is coming from a direction at right-angles, the vibrations induced across the array will be in phase, whereas from any other direction, they will be partially out of phase. As the echo intensity depends on the vector sum of the vibrations, it will be at a maximum when the target is in the centre of the beam. So, if a transducer is rotated across the bearing of a source of sound and the strength of the incoming signal is plotted against bearing, the result will be a polar diagram of the type already described. It follows, therefore, that a transducer has a bearing discrimination that is twice as good as the transmitted beam width. Suppose that it is of a size to transmit over a total angle of 10° to the 3dB points, then the response to a single target echo will be 6dB down 5° on either side of the bearing.

The assumption has been made so far that the amplitude of oscillations is uniform across the face of the transducer. The response can be varied, however, and the effect of this is to reduce, or enhance, the amplitude of the side lobes, either of which is desirable for certain purposes. This technique is known as tapering, or shading. Various methods have been adopted. For example, in a magnetostriction transducer a proportion of the laminations can be made of some inert material in place of nickel, and this proportion can increase towards the outer edges so that the transducer response is progressively diminished (see plate, p 34). Alternatively, inert laminations can be concentrated in the centre to enhance the secondaries at the expense of the main beam. Other methods are available with different types of transducer. The trade-in for a reduction in side lobes is an increase in the main beam width.

PROPAGATION LOSSES

The limitations of sound as a means of gathering information lie in the nature of the sea itself, which is far from being a homogeneous medium, and upon the speed of sound in water.[2] Sound propagation losses and distortions are grouped under the following headings:

Spreading
Reflection
Scattering
Absorption
Refraction

Each of these will be discussed in turn.

SPREADING

Spreading loss is independent of frequency and is inversely proportional to distance. For spherical waves the loss per distance doubled is 6dB. A significant reduction in spreading loss occurs in certain anomalous propagation conditions described under Refraction.

REFLECTION

Whenever the sound beam encounters a boundary between one substance and another, a proportion of the energy will be reflected and the remainder will penetrate into the new substance. The distribution of energy between these two paths is defined by the reflection coefficient originally calculated by Lord Rayleigh, and depends on the ratio of acoustic impedances of the two substances. In the case of air and water the ratio is 4,000 to 1 and virtually total reflection occurs, but the acoustic impedances of rubber and water are much the same and so sound will pass from one to the other with little loss. Reflection from the seabed is more complicated as there are interference effects due to sub-bottom layering, and also because the sound velocity in material on the bottom depends not only on density but also porosity, defined as the fraction of sediment volume occupied by water.

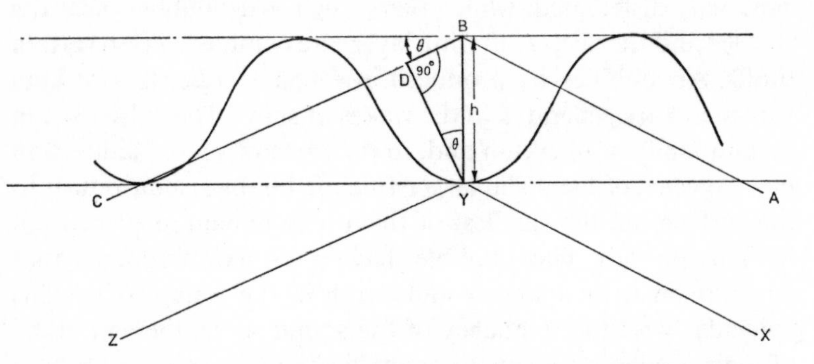

FIG 3 The typical amplitude of surface roughness is given by h. The difference in the reflected path distances from the top and the bottom of this surface irregularity is twice BD = 2h sinθ. The energy scattered on reflection decreases with θ. The value of θ when the scattering is reduced to zero is known as the grazing angle. The surface then acts as though it were smooth. This point is reached (according to Rayleigh) when the path difference (2DB) is less than $\lambda/4$, ie, when λ is greater than 8h sinθ

SCATTERING

The relationship between reflection and penetration holds good only for smooth surfaces. The surface of the sea is scarcely ever completely smooth and so, upon reflection, there is some energy loss through scattering, and this increases the rougher it is. However, 'roughness' in this context does not simply depend on sea state; it is also relative to the wavelength and the angle of incidence. As the latter is decreased, less energy is scattered until a point is reached when the boundary acts as though it were perfectly smooth and there is no scattering. This is known as the grazing angle.

In addition to the scattering of sound energy that occurs at both its upper and lower boundaries there are many inhomogeneities in the sea that have a scattering effect. They vary in size from particles of dust, minute air bubbles and micro-organisms to fish and fish shoals. Some of these scatterers are randomly distributed, while others, such as air bubbles near the surface and the deep scattering layer, are confined within certain limits. Air bubbles are produced in rough weather by breaking waves and are generated in the wakes of ships. They also exist in certain kinds of plankton and, in the form of swim bladders, in most species of fish. The larger free air bubbles soon return to the surface but the smallest of them may remain in suspension for long periods. These bubbles have a resonant frequency that depends on their diameter and depth in the water. When this coincides with the frequency of the sound wave, the amplitude of their oscillation is greatly magnified and more sound energy is scattered and lost in heat as a result. The diameter of a bubble that would resonate at 50kHz in a depth of 100ft (30m) is 0·013cm.

ABSORPTION

Absorption loss is the energy in the sound wave that is converted into heat by frictional effects occurring during the energy transfer process. The loss rate varies with the material involved and increases with frequency. The absorption rate in seawater is higher than predicted by basic theory. In the 5–50kHz region it is thirty times as high as it is in fresh water. An explanation for this has been found in the ionic relaxation of magnesium sulphate ($MgSO_4$), which, although only a trace element in seawater, is the dominant factor in absorption losses below 100kHz. The absence of this chemical in the waters of Loch Ness results in significantly better sonar performance than elsewhere. In the frequency band below 5kHz theory and practice are still not reconciled, and there appears to be a further, as yet undiscovered, process at work.

Scattering and absorption losses are sometimes collectively referred to as attenuation. Both are independent of range but increase with frequency.

REFRACTION

Refraction refers to the deflection of sound waves caused by the velocity gradients in the sea. In practice refraction effects have serious consequences in horizontal echoranging, but in echo-sounding, where the sound travels vertically, ie in the same direction in which the speed changes, there is no problem. The sea has a vertical sound velocity profile which, in the near surface regions, is continually varying. In consequence the upper and lower limbs of a horizontally directed sound beam will usually travel at different speeds, and the vertical section through the 'wave front', ie, the line of equal phase, will be tilted up or down and the whole beam bent accordingly. The sound beam can be visualised as consisting of a number of discrete rays. The direction followed by these rays is defined by Snell's law, which states that $\cos\theta/C$ is a constant, where θ is the

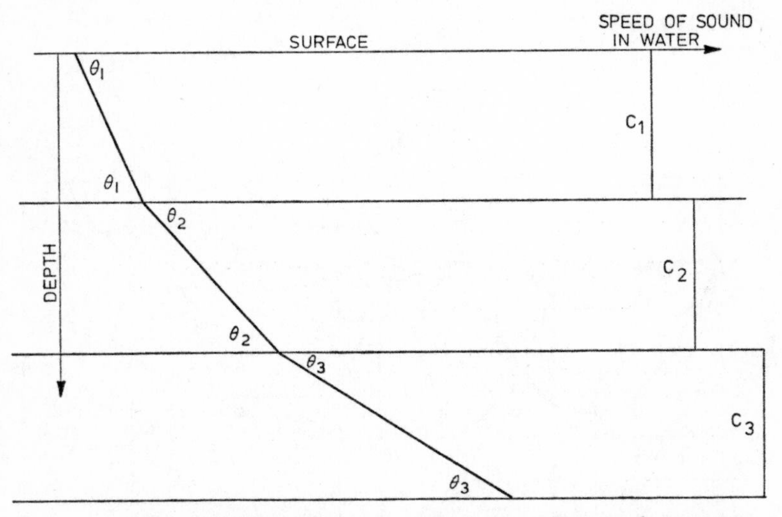

FIG 4 The path followed by a sound ray in a medium such as water in which the speed is changing is such that $\cos\theta/C$ remains constant. In the diagram $\cos\theta_1/C_1 = \cos\theta_2/C_2 = \cos\theta_3/C_3$. If the sound velocity is changing continuously at a uniform rate, instead of in discrete steps as shown, the ray path becomes a curve in the form of a cycloid

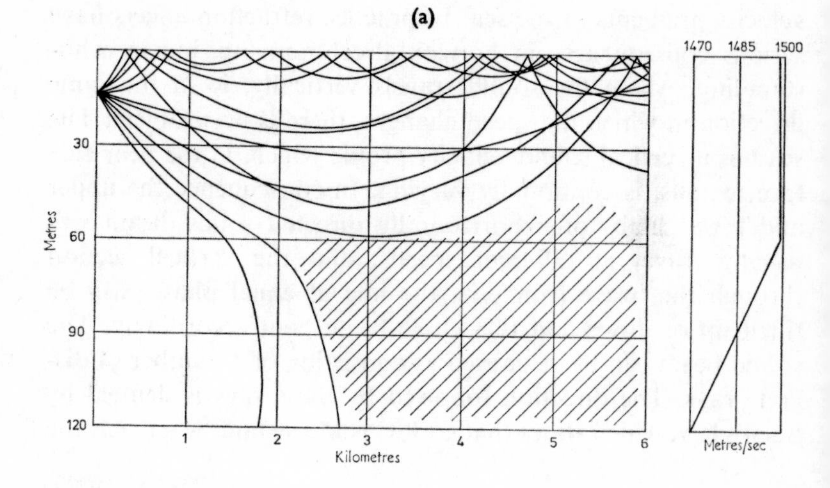
(a)
1470
1485
1500
30
60
90
120
Metres
1
2
3
4
5
6
Kilometres
Metres/sec

(b)

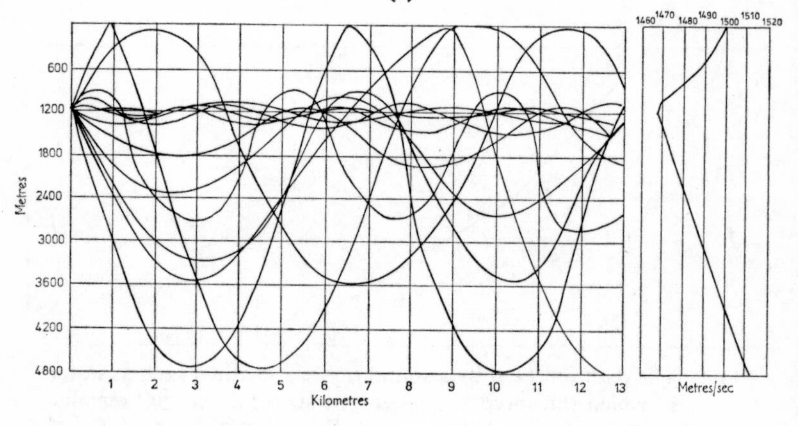
1460
1470
1480
1490
1500
1510
1520
600
1200
1800
2400
3000
3600
4200
4800
Metres
1
2
3
4
5
6
7
8
9
10
11
12
13
Kilometres
Metres/sec

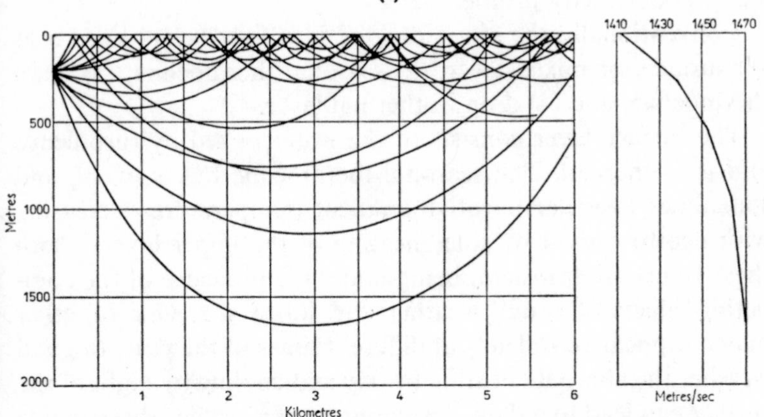

FIG 5 The diagrams show typical ray paths for various sound velocity profiles, shown on the right of the ray diagram in each case. Fig 5a is an example of sound ray paths in normally turbulent sea conditions with a mixed, and therefore isothermal, layer extended to 60m. Sound speed slightly increases with depth in this zone owing to the effect of density. Echo detection in the shaded area is poor, though not non-existent, owing to the presence of volume scatterers in the water. Fig 5b illustrates the effect of placing a sound source in the deep sound channel. At this depth the competing effects of increasing pressure and a negative temperature gradient have a focusing effect on the sound speed profile. It will be seen that the ray path down the centre of the channel is travelling at the slowest speed. Although it has the shortest distance to go, in fact it arrives last. The sound, initially of short duration, becomes spread out with range and is received as a long crescendo ending with a sharp cut-off after the arrival of the sound by the direct path. This enables the time to be measured with considerable accuracy. Fig 5c shows the effect of a sharp positive temperature gradient such as is found in water under ice (after R. J. Urick. *Principles of Underwater Sound for Engineers*)

angle of incidence and C the speed of sound (Fig 4). Ray path analysis is the method used to predict sonar performance from the sound velocity profile.

Conventionally the sea is divided into four layers known as (1) surface, or mixed layer, (2) seasonal thermocline, (3) main thermocline, and (4) deep isothermal layer.

The surface layer consists of the water mixed by turbulence and is isothermal. The seasonal thermocline has a strong and sometimes irregular negative gradient (temperature decreasing with depth) caused by solar heating of the upper layers. Both these layers are transient, being under the influence of the competing effects of rough weather and warm sun. One or other may disappear completely at different times of the year. In good weather the alternate heating by day and cooling by night of the surface can lead to a daily development of a surface thermocline resulting in noticeably poorer sonar performance. This has been called the 'afternoon effect'.

In the main thermocline there is a steady negative temperature gradient below which, in the deep isothermal layer, the temperature remains constant, and the sound speed under the influence of the increasing pressure slowly increases with depth.

Between these two there is a minimum value for the speed of sound, and this has a focusing effect on the ray paths. This is known as the deep sound channel in which, because of the elimination of spreading loss, sound from a source at this depth can be detected after travelling well over 1,000 miles. This is a case in which the shortest path takes the longest time.

Ducts or sound channels of a less permanent nature can also occur in the mixed surface layer and in shallow water when the sound is trapped between surface and seabed. However, in these cases losses due to scattering on reflection are considerable. Special conditions exist under the ice, where there is a positive temperature gradient bending the sound rays sharply upwards. Some examples of ray paths under various conditions are shown in Fig 5 and the plate on p 51.

BACKGROUND NOISE

The background noise against which echoes must be detected varies greatly but is always too high for the electrical noise of the receiver to become a limiting factor in performance, as is the case, for example, with radar. Noise from the sea falls into three general categories—ambient noise, self-made noise and reverberations—each of which will be considered in turn.

Ambient noise comes from many sources, not all of which have been fully analysed. Starting at the low frequency end, microseismic activity with a period of 7sec is a source of low frequency sound, and there is some evidence that turbulence caused by deep ocean currents is a contributory factor. There are enough ships at sea at any time to create a general background noise with a dominant frequency of 100Hz, and distant storms add to this. Surface waves and wind on the surface cause noise in the near region from 500Hz to 30kHz. In coastal waters tidal streams cause noise on the seabed particularly when it is composed of rough or unstable material, and sound from this source has been recorded at frequencies as high as 300kHz. Marine life in the sea is responsible for a good deal of noise, sometimes sufficiently loud and continuous to limit sonar performance. Notable in this category are the snapping shrimps found off the coast of California and elsewhere, which cover a wide band of frequencies, from 500Hz to 25kHz. A few species of fish, such as the croaker, which makes a noise like a woodpecker, can be detected by the sounds they make; but unfortunately this does not apply to any of the species that are sought after commercially. The toothed marine mammals, including sperm whales, killer whales, dolphins and porpoises, all use sound signals to communicate with each other and for echolocation.

Self-made noise includes all the various sources of noise emanating from one's own ship. Foremost among these is the propeller noise, which together with cavitational effects greatly reduces sonar performance while under way anywhere within about 15° of the astern bearing. The spectrum of this sound is 'white' as far as 30–50kHz and above that on a diminishing scale. Flow noise of water past the hull is considerable at speed, and, in rough weather, quenching and bubble sweepdown can mask the transmission completely. When transmitting horizontally with the face of the transducer in the vertical plane, at speeds above 10–12 knots (the exact figure depends on the ship), bubbles begin to form and collapse, causing noise.[7] This flow cavitation effect builds up rapidly, but its onset can be delayed by encasing the transducer in a streamlined dome with an acoustically transparent envelope. Reasonable sonar performance can then be achieved up to ship's speeds of 20–25 knots. The downward looking echosounding transducer is less affected by this form of cavitation, provided that it is properly sited in the hull. Good results have been achieved in high speed patrol craft at between 40 and 50 knots.

When the negative acoustic pressure at the face of the transducer exceeds the hydrostatic pressure, bubbles begin to form and collapse, causing noise and loss of efficiency. This condition, known as acoustic cavitation, sets an upper limit on the amount of power that can be transmitted per unit area. Of course, the deeper the transducer, the higher this limit becomes. For steady state conditions and at atmospheric pressure it has been put as low as 0·3 watts/cm^2 for the 10–100kHz range. However, the condition does not develop immediately, and somewhat higher powers are possible when using short pulses.

Noise from machinery inboard, particularly flow noise from hydraulic equipment, can be transmitted via the hull to cause acoustic interference.

The third type of background noise, known as reverberation, is only loosely associated with the other two, since it is of an essentially different nature. Reverberation is a direct result of the transmission itself. The many types of scatterers in the sea have already been mentioned. They not only attenuate the sound pulse but also reflect back to the transmitter a small part of the energy, which is picked up in the form of an undulating series of weak overlapping echoes. As each of these individual echoes is of exactly the same form as the echo from the target being searched for, though of a lower amplitude, they cannot be designed out of existence in the amplifier. Nor is it possible to reduce the effect relative to the target echo by increasing the power of the transmission, as this would only raise the amplitude of both types of echo proportionately. Reverberation decreases exponentially with range, but more slowly than the target echo, because, as the beam becomes wider, more scatterers fall within its scope. The maximum echo-detection range is said to be reverberation-limited or noise-limited, depending upon the relative amplitude of noise, reverberation and echo (Fig 6).

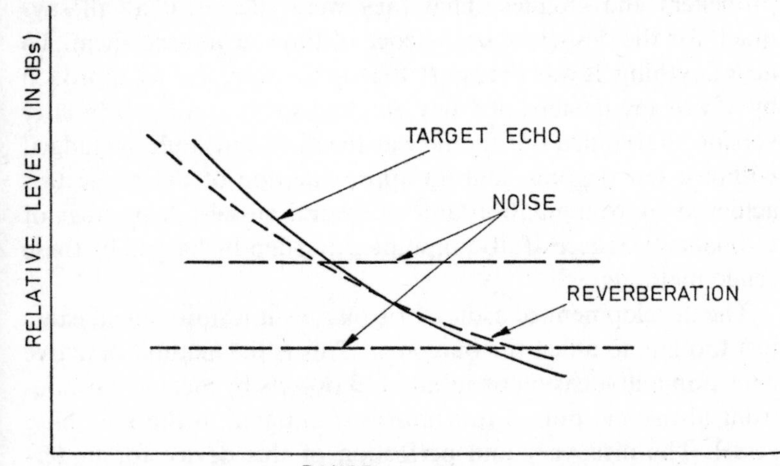

FIG 6 The volume reverberation echo (dotted line) decays more slowly with range than the target echo because attenuation is reduced by the inclusion of more scatterers in the beam as the range increases. The graph shows two levels of background noise. When the background noise level is high, in the case shown, this determines the maximum range at which the target can be detected. It is then said to be 'noise limited'. Alternatively, when the background noise is low, the target is 'reverberation limited'

Notes to this chapter are on page 191

2

ANTI-SUBMARINE WARFARE

IN WORLD WAR I SUBMARINES COULD ONLY BE DETECTED BY passive means when submerged, ie, by sonic listening devices or hydrophones. These gave no indication of range and, to begin with, virtually none of bearing, and were wholly dependent for their effectiveness upon the sounds made by the submarine's propellers and engines. That they were successful at all says much for the desperate persistence of those who used them. To hear anything, it was necessary to stop the ship, and to approach by advancing in steps of a few hundred yards at a time. In later versions developed during the war the direction could be judged within a few degrees, and an approximation of the range was achieved by triangulation among several vessels. A number of U-boats were successfully pinpointed and depth-charged by these crude methods.

The development of asdics, or sonar, as it is now called, came just too late to affect the outcome.[1] This is the method of active detection and location of submerged objects by means of echoes from ultrasonic pulsed transmissions initiated in the searching vessel. The discovery and perfection of this device for underwater echolocation has some interesting parallels with its much better known counterpart in the air—radar.

Both systems proved to be of vital importance in World War II, the former in the Battle of the Atlantic and the latter in the Battle of Britain. Both, too, have since been adapted for commercial purposes; but whereas the story of radar has oft been told, the history of sonar, in its own way just as dramatic, seems never to have percolated beyond the restricted circles of those

Page 33 *The first scientist in ultrasonics*. In this photograph taken at Toulon in 1918 Prof Paul Langevin is the man in the bowler hat. Prof R. W. Boyle, who was leading the British research team at that time, is in the centre

Page 34 *Magnetostrictive Transducers.* (*left*) An early toroidally wound ring type. The larger unit was designed for 16kHz. (*below right*) A fish detection sonar transducer. The main face is tapered in the vertical plane, the lighter coloured stampings being of an inert material. (*below left*) A recent design consisting of 121 separate elements. This transducer is mechanically stabilised and tiltable. It produces groups of fan-shaped beams simultaneously

directly concerned. This can be accounted for by the accident of timing. The discovery of the magnetron, at Manchester University, which paved the way for the production of centimetric radar, occurred shortly before the outbreak of World War II, but the comparable breakthrough in the development of sonar occurred towards the end of World War I. In consequence the cloak of secrecy that surrounded the original work was maintained throughout the interwar years, and was more complete for the fact that sonar was only used in action for the first time in 1939, so that its very existence was hitherto unknown outside the groups of scientists and their associated naval colleagues in both Britain and America who were working on it. Even quartz, used in the construction of the transducer, was invariably referred to as Asdivite to put the uninitiated off the scent.

So there was a lapse of over 20 years between the first successes with sonar and its apotheosis in the Battle of the Atlantic, itself a long drawn out struggle which took 3½ years to reach its crisis, and then continued on a slowly diminishing scale for the rest of the war. Several more years elapsed before the wartime designs were declassified. By then the early history of the development lay too far in the past to excite much general interest and, in any case, many of the original papers had long since disappeared.

For the benefit of any who still doubt the significance of this detection system in mastering the threat of submarine warfare, which came, in two world wars, within measurable distance of achieving decisive results, the words of a directive issued by Admiral Doenitz in 1943 will suffice:

> For some months past the enemy has rendered the U-boat ineffective. He has achieved this object, not through his superior tactics or strategy but through his superiority in the field of science; this finds its expression in the modern battle weapon—detection. By this means he has torn our sole offensive weapon in the war against the Anglo-Saxons from our hands. It is essential to victory that we make good

our scientific disparity and thereby restore to the U-boat its fighting qualities.

The early development of this ultimately decisive 'weapon' took place in a world in which science and technology had yet to make its impact. During World War I and in the years immediately following there was little general interest in R & D and, generally speaking, much of the initial work was carried on under the handicap of limited material support and meagre facilities. Necessity, too, forced the scientists concerned to develop a technical versatility unheard of today. There was no body of research to fall back on and almost everything had to be thought out and made from scratch, a point which may be emphasised by the fact that up to the middle of 1917 less than one merchant ship in five entering and leaving British and French ports carried any wireless equipment at all. For navigation, the lead and line, magnetic compass and sextant sufficed. It is against this background that the pioneering work of those days must be considered.

The possibility of the echolocation of objects under water was first thought of in the aftermath of the loss of the *Titanic* after striking an iceberg in the Western Atlantic during her maiden voyage in 1912. L. F. Richardson patented a proposal for an ultrasonic detection device in the same year, but never pursued it.[2] The American scientist R. A. Fessenden went one further, and on 27 April 1914 detected an iceberg by echoranging with his moving coil transducer at nearly 2 miles (3km).[3] Presumably, however, this device was non-directional, as it operated at the low frequency of about 1kHz.

It was at about this time that the eminent French physicist Paul Langevin began to study the problem of ultrasonic echolocation. He was joined by the brilliant and volatile young Russian, Constantin Chilowsky, who came to him from a sanatorium in Davos, where he had been recovering from TB, and the two of them carried out a series of experiments using the piezoelectric effect of quartz. Working first in Paris and later at Toulon, they made rapid progress. They first succeeded in initiating vibrations at 100kHz with mica used as a dielectric subjected to electrostatic stresses; and for receiving the signals they used a microphone and tuned resonant circuit. By 1916 signals had been recorded at a range of over 2km, and echoes from the seabed and a sheet of armour plate at 200m. During this time Langevin maintained a close liaison with Sir Ernest (later Lord) Rutherford and through him with the Canadian, Prof R. W. Boyle, who led the British team that was working on this problem and provided the driving force.

The possibilities of quartz, at that time still only used as a hydrophone, was the subject of a memorandum from Rutherford to the Admiralty; there was great promise in the development, he said, 'but it is a pity that the effect is so small'. This was before the development of high gain amplifiers.

Langevin was the first to try using a plate of quartz as a receiver in conjunction with a new amplifier just developed by l'Ècole Centrale de T.S.F. This represented a big step forward but the real breakthrough came with his quartz sandwich design consisting of a thin sheet of quartz between two steel plates, the total thickness determining the resonance. Using this device as a transducer, Langevin succeeded in obtaining echoes from a submarine at a range of 2–3km.

Quartz crystals of a suitable size suddenly became as valuable as uranium in World World II, and large natural crystals in geological museums and jewellers' shops were quickly tracked down on both sides of the Channel. Boyle discovered a bonanza in a Bordeaux warehouse. The bottom, it seemed, had fallen out of the domestic market.

The problems of slicing the natural quartz into the most efficient p-e slabs, and the mounting of the slices into mosaics, all had to be worked out theoretically and experimentally in short time. In Britain the cutting was done by Farmer & Brindlay, the tombstone-makers of Lambeth.

ANTI-SUBMARINE WARFARE

To assess the status of the ultrasonic developments of U-boat detection devices in the four Allied countries principally concerned—Britain, France, Italy and the United States—an inter-allied conference was held in Paris from 19 to 22 October 1918. Each country sent four delegates. At that time the British and French were working on closely parallel lines. A report on work carried out in the United States since its entry into the war and presented to the conference is still extant, but other papers have not been traced.

The end of the war, less than a month after this conference, came before any of the ultrasonic developments that were being so actively pursued had reached fruition. It brought collaboration to a halt, and thereafter the countries concerned went their own way. Small British and American teams were kept in being, and continued to work on the anti-submarine detection problem, though inevitably at a much reduced tempo. The French did little more in the naval field but, instead, adapted Langevin's research for hydrographic and navigational purposes, ie, in the development of echosounders rather than sonar systems. This is discussed further in Chapter 3.

BRITISH DEVELOPMENTS, 1915–39

In 1915 the Admiralty Board of Invention and Research (BIR) was founded and became a landmark as the first formal link between the scientist and the Navy. The central committee contained many famous names, among them Lord Fisher, who was the first president, J. J. Thompson and Lord Rutherford. Section II of the organisation was concerned with submarines, mines, searchlights, wireless telegraphy and general electrical, electromagnetic, optical and acoustic subjects. Among this wide list the anti-submarine problem had the top priority. Dr A. B. Wood, who had previously been working in Rutherford's laboratory in Manchester, was the first physicist to receive an Admiralty appointment under BIR and thus became the first member of what was much later (in 1944) to become the Royal Naval Scientific Service (RNSS), now numbering around 2,000. He spent the whole of his working life serving the Admiralty in many capacities, much of his time being devoted to underwater sound. Along the way 'AB' gained many honours and distinctions and in 1930 published *A Textbook of Sound*, the definitive work on the subject—the 'ABC', if one may presume to call it so, of acoustics.

Most of the early Admiralty experiments involved the use of moored, portable and hull-fitted hydrophones. The propeller noise of the ships of Beatty's Battle Squadron based at Rosyth could be heard at a distance of 10–12 miles (16–19km) under favourable conditions. While the sounds came essentially from the propellers, these were modulated by the engines: the beat of a drifter's reciprocating engine could be distinguished from the more continuous sound from the turbines of destroyers and larger warships. These hydrophones were completely non-directional, but by selecting pairs with different diaphragm resonances and fitting them one on each side of the submarine's bow, and taking advantage of the masking effect of the hull, it was found possible with skill to obtain some idea of the bearing of a source of sound.

A. B. Wood and his associates tested various types of sound receivers for use in the sea. One was the 'Broca' tube, essentially Leonardo da Vinci's sounding tube with a rubber diaphragm at the bottom and a stethoscope at the top.

The first, not unsuccessful, attempts were made at this time to design a portable directional hydrophone. A microphone was mounted in a small watertight box at the centre of a diaphragm surrounded by a heavy metal rim. With both sides of the diaphragm exposed to the sea, the device gave a reasonably good figure of eight response with an almost silent minimum when the diaphragm was edge-on to the sound source. A double diaphragm design gave an equally good uni-directional response with an accuracy of one to two degrees. Experiments with towed hydrophones were unpromising due to excessive water noise and vibration of the towing rope.

In 1917 a series of experiments using sea lions for submarine detection was mounted and duly reported on by BIR. The conclusion reached was that 'it is recommended that these animals should now be allowed to return to their legitimate business'.

Work on hydrophones was continued at the Admiralty Experimental Station at Parkeston Quay, where a whole variety of designs were made. They were used in pairs or groups with the output connected one half to each ear, the whole array being trained for equal response to give bearing. This binaural technique was reasonably effective. As an alternative to mechanical rotation of the array, an American device known as a compensator was used; this introduced a variable delay into one side of the array, and the bearing of the noise source was obtained by adjusting the delay until both sides were matched.

Hydrophones were fitted to small craft in considerable numbers from mid-1917 until the end of World War I. Sea and propeller noise limited their effectiveness, and they were more valuable in submarines, where the ambient noise was at a lower level when submerged and particularly when lying stationary on the bottom.

Calibration work with these hydrophones led to the idea of an acoustic non-contact mine, and this was shown to be a practical proposition at Parkeston Quay in May 1917. Development trials were started in the autumn of that year and continued until April 1918. Several thousands of these mines were ready for laying when the war ended in November 1918. They did not survive the peace, but an improvement to A. B. Wood's contact firing mechanism, known as the 'shunt-relay' acoustic firing system, was in fact used extensively and successfully in World War II.

The acoustic mine was by no means the only development in those early days that had its counterpart in World War II a quarter of a century later. Another example was the work begun at Parkeston and continued after the end of the war by the Admiralty Research Laboratory (ARL), Teddington, on underwater sound ranging. This method was akin to the American technique for determining the positions of ditched aircraft, known as SOFAR (for Sound Fixing and Ranging) during World War II, though without benefit of the deep sound channel, then unknown.

The idea for the earlier sound-ranging system came from an army method of locating enemy guns and involved the recording of the explosion (or gun blast) by a series of hydrophones mounted at fixed positions along a surveyed base line. The hydrophone responses were recorded on an Eindhoven 'string' galvanometer, and it was found that even a small charge gave a positive break in the record, which could be timed with considerable precision. A detonator could be recorded at 2–3 miles and larger charges at correspondingly greater ranges. The speed of sound in water was assumed to be constant and initial trials at Culver in the Isle of Wight included a careful determination of its value; observations over several months confirmed that changes due to temperature were within acceptable limits. The accuracy of the system depended on the spread of the recording hydrophones.

The initial investigations were sufficiently promising to persuade the Admiralty to set up operating stations at St Margaret's Bay near Dover, at Easton Broad near Southwold in Suffolk, at Flamborough and at Peterhead. These four stations, acting singly or together, were able to keep watch over a large area of the North Sea, especially the southern stretch along the Dutch coast. The base line at St Margaret's Bay was in a north-south direction, extended for about 70,000ft (20km) and was made up of five hydrophones. The time intervals between the arrival of an explosion pulse at each of the hydrophones could be estimated with an accuracy of 1ms. The galvanometer responses were recorded photographically on a continuous strip of bromide paper, over a mile a day, moving at about 1in per sec, which could be inspected within a minute or two of exposure after developing and fixing. A continuous watch therefore involved the inspection of huge lengths of useless record, and it was not long before the designer devised a relay system to start the recorder on receipt of an explosion signal from the nearest hydrophone and to switch it off again after all the pulses had been received. The stations at St Margaret's Bay and Southwold were successfully used to fix the positions of minefields laid off the Belgian coast 20–30 miles away, with an accuracy of 100yd.

The St Margaret's Bay station was kept in operation after World War I, and was used for further experimental work in submarine sound ranging and for measuring the effect of temperature and salinity on the speed of sound.

But it was the initial work on sonar, in parallel with the Langevin/Chilowsky experiments, that was subsequently to prove the most successful and significant of all the acoustic work at AES, Parkeston. This work was under the direction of Professor Boyle, and two members of his research team, B. S. Smith and J. Anderson, each in turn became Chief Scientists of the Admiralty Underwater Detection Establishment (UDE), covering between them a span of 35 years and exercising a profound influence on the whole technology of submarine detection. For his research work and collaborative efforts during World War II, Jock Anderson was awarded the Medal of Freedom by the US Government in 1947.

In trials from the drifter *Hiedra* in the estuary of the River Stour a 700 ton vessel was detected at 1,400yd (1·3km) and similar ranges were obtained with a submarine on the surface. The submerged detection range was a good deal less than this. These results, remarkable for those days, were achieved early in 1918; but the war ended as the experimental phase was giving way to development and ship fitting, and there was then, inevitably, a slowing down of the whole programme. In 1919 a quartz transducer was first tried with a circular dome in the auxiliary craft P59. The next year the group was moved to Portsmouth and finally, in 1927, the Underwater Detection Establishment, as it then became, was set up at Portland, where it has remained to this day. HMS *Osprey*, the school founded in 1920 for the training of officers and ratings specialising in anti-submarine warfare, was half a mile away up the hill, perched on the edge of the massive Portland Bill rock.

During the years of consolidation, 1921–39, the principles of active detection, or echoranging, on submarines were proved and the design of equipment gradually improved. Many problems were encountered and solved. One of these was to discover a suitable cement for jointing the parts of the quartz mosaic without inhibiting the oscillation of the transducer face as a whole. Another was concerned with flow noise caused by turbulence in rough weather and by the movement of the ship through the water. A cylindrical housing is effective up to about 10 knots, but above that speed bubbles begin to form and collapse in the housing. The onset of this form of cavitation can be delayed by using a streamlined shape. But what was the best shape? There was not then, as there is today, an extensive literature on the subject, and though much was known about streamlining from the point of resistance, flow noise was a new subject. To find out about it, an observation dome was mounted on the cruiser HMS *Devonshire*—with scuttles, and nozzles for measuring pressure—20ft (6m) from the bow. Readings of pressure all round and visual observations of aeration were made at speeds up to 30 knots by a small and intrepid member of the research team, Jock Anderson. He said it was hair-raising. Out of this work came the streamlined shape with a length/breadth ratio of about 3:1 which enabled operations to be continued at speeds up to about 20 knots, and which was, in fact, the most important difference between the British and American sets as they were in 1939.

In the meantime, a ship-fitting programme was begun. First the A/S training flotilla and later fleet destroyers and submarines were fitted with sonar, until by 1934 all ships on active service from cruisers downwards were equipped, and installations in the reserve fleet of V & W destroyers, dating from World War I, in mothballs at Rosyth had begun.

Up the hill, training methods and tactics were studied and developed and the procedures for classifying echoes into the immemorial categories 'doubtful', 'sub' and 'non-sub' were worked at. Great store was set by aural acuity. The doppler shift in the pitch of the target echo caused by a relative movement along the 'line of sight' was always the most telltale indication that the echo under investigation was the real thing, but all too often the wish was father to the thought. Those with an infallible sense of pitch discrimination were at a premium, but those with only a good imagination were not.

US DEVELOPMENTS, 1917–39

The early development of hydrophones for use in the US Navy followed broadly similar lines to the British programme. The 'Broca' tube, already mentioned, was developed and fitted in both submarines and patrol vessels. It was later expanded into a system containing six or more nipples on each bar; and in turn replaced by an electrical sonic system, consisting of a blister on each bow containing twelve microphones of the carbon button type that fed the sounds through electrical phase lines to a compensator to give direction to within a few degrees.

Shortly after America's entry into World War I the recently formed National Research Council (NRC) organised a conference in Washington, DC, at which the British and French developments were described. The US contingent included representatives of GEC, Western Electric and Subsig, who were already carrying out joint experiments at Nahant, Massachusetts.

After the war's end, work continued on a greatly reduced scale. The NRL sound division moved into new laboratories at Washington, DC, where it remained throughout the interwar years under Dr Harvey C. Hayes. It is remarkable how much this group achieved, considering how small it was: in 1927 there were exactly five scientists (including Hayes) on the staff of what was the US Navy's only research and development centre devoted to underwater acoustics. Moreover, in the atmosphere of the Washington Naval Conference, it was inevitable that the programme was starved of funds. As one writer expressed it: 'If it were possible and popular to sink battleships, it was even easier to wreck a research programme.'

The smaller vessels of the US Navy were gradually equipped with echoranging instruments of progressively more advanced design. The first sets used the quartz/steel sandwich type of transducer and operated at frequencies between 20kHz and 40kHz. The more sensitive Rochelle salt crystals were used in place of quartz for the submarine listening set, known as 'JK',[4] which gave ranges of 5 miles on moving targets and bearing accuracy within a few degrees. Later modifications included a low power transmitting element for communication between submarines. An echoranging set, QB, was developed by NRL in 1931 and manufactured by the Submarine Signal Company. Later, the quartz/steel type of transducer was superseded by a new design using the magnetostrictive principle, and it was with this that the US Navy was equipped at the outbreak of World War II. These sets had a spherical rubber housing for the transducer, limiting effective operations to about 10 knots, and it was the streamlined transducer housing developed in Britain that represented perhaps the most important operational difference between the British and American equipments at the outbreak of World War II. Ironically however, it seems to have been an American who first proposed the idea of a streamlined housing for the transducer. In August 1918 the US Scientific Attaché in Rome, when reporting on the Langevin experiments at Toulon, introduced the idea of fitting a sheet steel streamlined cover, $\frac{1}{16}$in thick, about the Langevin device to reduce the difficulties of steering it in the ship while under way.

The other main difference between the two types of equipment was in the method of displaying the echo information. Both sets had an audio output, which enabled the ultrasonic echoes picked up by the transducer to be heard, either by loudspeaker or telephones, after heterodyning to a sonic frequency around 1kHz; but in the American sets the range of the echo was indicated on a dial, whereas the British ones had a chemical recorder of the type used for echosounding (described in Chapter 3). This had the advantage of showing the rate of approach, or range-rate, of the target, from which the time to fire depth charges or, later on, ahead-thrown weapons, could be calculated.

At the outbreak of World War II sonar was already fitted to 165 British destroyers, thirty-four patrol craft and twenty trawlers. In the United States about sixty destroyers were fitted.

WORLD WAR II

The sonar systems developed and brought into production in Britain and America between the wars and the tactics worked out for their use stood up to the test in World War II extremely well. Perhaps inevitably the persistence needed to secure a kill was underrated. Peacetime exercises for all sorts of administrative reasons last only for a predetermined time, whereas the real encounter is open-ended. In the earlier months of the Battle of the Atlantic a good many U-boats escaped because the action was broken off prematurely. Some extraordinary cases of survival against all the odds were recorded. The ability of the welded hulls of the German submarines to withstand shock had probably been underestimated, and wartime experience indicated that the hydrostatically operated pistol in the depth-charge firing mechanism was less accurate than had been assumed. Nor were the U-boats slow to think of ways of persuading their attackers that the battle was over. Floating debris of all kinds was ejected, together with quantities of diesel fuel. But as the war progressed such 'evidence' of a kill soon became suspect and, in the absence of survivors, something more grisly than pieces of clothing was needed for conclusive proof. A human ear in somewhat decrepit condition, packed in a small box was said to have reached the Admiralty on one occasion, by way of supporting evidence.

The submarine is aided by the third dimension, depth, and in the first types of sonar there was no method provided for measuring the depth of the target below the surface. As the ship approached for the attack, a time would come when the target echo would be lost as the submarine came in below the beam. The range at which this occurred provided a first approximation of the submarine's depth in sufficient time to set the depth charges accordingly. The deeper the submarine went the greater the opportunity for error and the longer the 'dead time' that elapsed between the moment of losing contact and the moment of firing, during which the submarine's alterations of course, speed and depth would pass undetected.

This problem of depth finding has its analogy in fish detection procedures, when it is a question of setting a mid-water trawl at the correct depth, but in this case the estimate can be checked by the echosounder as the ship passes over the shoal. In the submarine attack it is by then too late, as that is the moment at which the charges are released over the side. In any case, the ship does not, or should not if the approach has been judged correctly, pass exactly over the target but slightly ahead of it. The few seconds taken by the depth charges to sink to their set depths are equated to the time taken by the submarine to reach the ship's track. On hearing an increase in propeller noise, and a more rapid rate of sonar transmission, a submarine, knowing that an attack was impending, would go deep and delay his escape manoeuvre to the last minute.

The interrelated problems of depth measurement and dead time were overcome by the introduction of a subsidiary sonar for measuring the target depth. In this the circular transducer face was replaced by one that was long and narrow; this so-called 'sword' was lowered beneath the hull to point vertically downwards and produce a sound beam that was broad in the horizontal plane and narrow in the vertical, like a fan. By tilting the sword about its upper extremity the fan-shaped beam would be swept up and down and so record the vertical angle of the 'line of sight' to the target, from which, knowing the range, its depth could easily be calculated.

In their endeavours to escape the attentions of escort vessels the Germans developed a 'Pillenwerfer' or 'Submarine Bubble Target'. This consisted of a canister of something equivalent to Eno's fruit salts which was pushed through the signal ejector tube to create a patch of bubbly water that returned a sharp echo very similar to a submarine's and subtending about the same angle. It would gradually expand and disperse after some minutes, but it retained its realistic character long enough for the submarine to break the contact and make its escape. It was effective initially but the measure was soon reported, and after that its value waned. The main limitation of the device was its inability to simulate movement, and when successive echoes, plotted for range and bearing, all seemed to come from the same spot, suspicion was at once engendered. As the SBT could not move, it could not exhibit the tell-tale doppler shift, and so was inevitably suspect.[5] As we shall see, the SBT has an exact equivalent in nature in the blow wake left behind in the water each time a whale surfaces to take a breath of air. This is equally effective, as it happens, in fooling the unwary operator.

The perfection by Germany in 1943 of an acoustic homing torpedo gave the U-boat for the first time a weapon that was effective against the A/S vessels as well as the merchant ships that were its main target. Inevitably some months elapsed after its first appearance before its full capability was known and effective measures to compete with it introduced. It was originally believed that an acoustic homing device fitted to a torpedo could operate effectively only at relatively low speed because of the nearness and noisiness of its own propeller. However, it turned out that the German 'Gnat' had a speed of 23½ knots, a remarkable achievement and one of considerable tactical significance, especially where attacks on escort vessels were concerned.

The antidote was a noise-maker, known as 'Foxer' which was towed on the quarter, but sufficiently far astern to ensure that it would not merely lead the torpedo to the ship. Combined with tactical avoidance manoeuvres during the approach, this was effective in mastering the new weapon.

The decisive convoy battles were fought in 1943 and thereafter the U-boat became steadily less dangerous and more and more on the defensive. The landing of the Allied Armies in Normandy in 1944 was scarcely affected by submarine attack.

When operating over the continental shelf, the water is shallow enough for submarines to rest on the seabed, where they are very difficult to locate. In the Irish Sea and the English Channel there are thousands of wrecks whose positions, in those days before the development of radio position fixing systems, were imprecisely known. A wreck and a live submarine give indistinguishable echoes and so when in doubt the only way to identify one from the other was to depth-charge it and sample the oil that came to the surface. If it were diesel oil, there was a probability—no certainty—that the target was a 'sub'.

In studying this problem of definition a scientist at UDE first hit on the idea of the acoustic shadow, which involves the transmission of a pulse of sound narrow in the horizontal plane and wide in the vertical (the opposite of the 'sword' depth-finding asdic) on a fixed bearing at right-angles to the ship's movement through the water. From Fig 7 it will be apparent that whenever there are objects, be they rocks, wrecks, submarines or anything else, that stand up above the floor of the seabed, there will be a gap beyond them on the acoustic record which is equivalent to the time taken by a pulse of sound to travel from the highest point of the object to the sea floor beyond. This gap, or acoustic shadow, is thus a function of the height of the object above the bottom. As the ship moves forward successive echo traces are produced on the chart, one below the other, which build up a profile of whatever is on the seabed that is causing the obstruction, and this profile makes it possible to identify one type of ship from another. The device has now become one of the most important tools in seabed mapping and hydrographic surveying and is widely used all over the world.

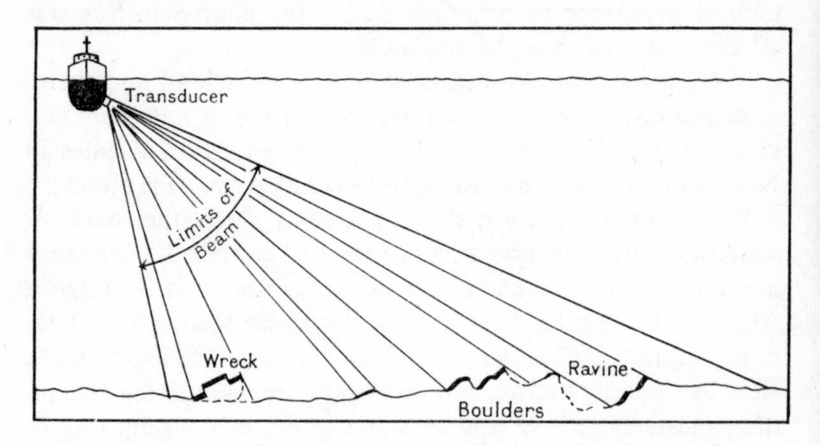

FIG 7 From this diagrammatic cross-section through the sidescan sonar beam it will be evident that the areas beyond the wreck and the boulders, and the nearer side of the ravine will be in shadow from the transducer. The 'echo time' for the sound to cross the gap in each case will appear on the chart as a blank space. As the ship moves forward, the lengths of these gaps will vary with each new transmission according to the shape of the object casting the shadow. A succession of such echo traces, laid one above the other on the chart, draw a profile of the object in this way which serves to identify it. The parts of the seabed that are more or less at right-angles to the wavefront (shown as a thicker line) will generally return a stronger echo. It will be seen that this comes *before* the shadow in the case of an object standing above the sea floor, and *after* it in the case of a dip or ravine. For good results the beam must be narrow in the plane at right-angles to the paper, otherwise some part of the sound pulse may reach the shadowed areas round the sides of the upstanding objects, thus blurring the record

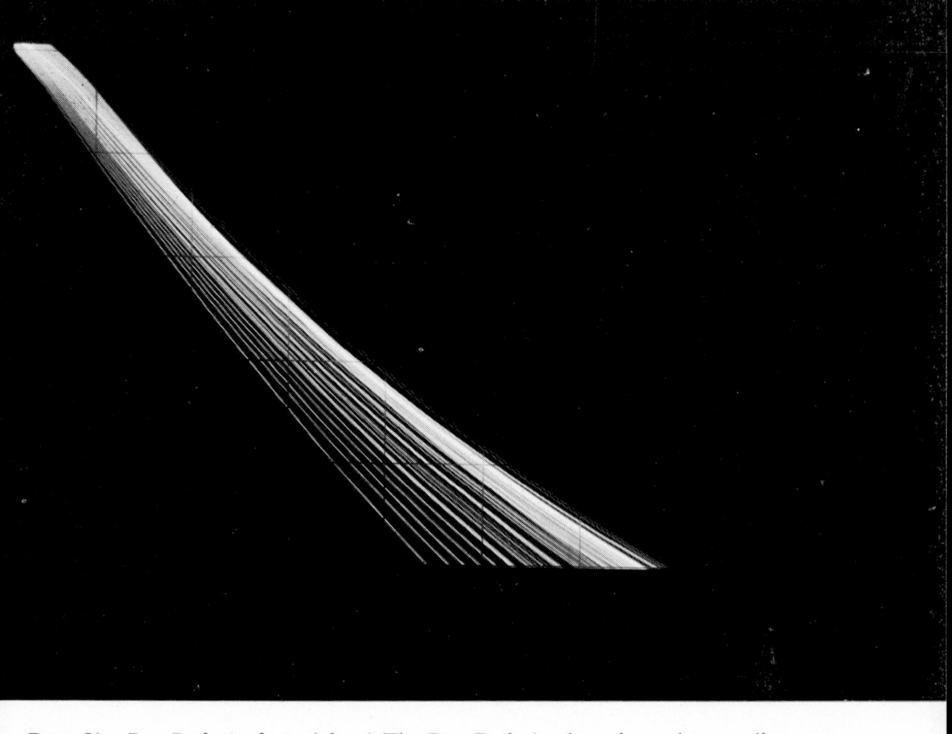

Page 51 *Ray Path Analysis*. (*above*) The Ray Path Analyser is used to predict sonar performance on the basis of local sound velocity profile. In this case the downward bending of the sound beam indicates poor performance in detecting mid-water and near-surface targets, with hull-mounted sonar. This is caused by strong negative temperature gradient (temperature decreasing with depth) typical of persistently calm and sunny weather. (*below*) This shows improvement in performance can be achieved under the same conditions as above by lowering the transducer in the water (VDS)

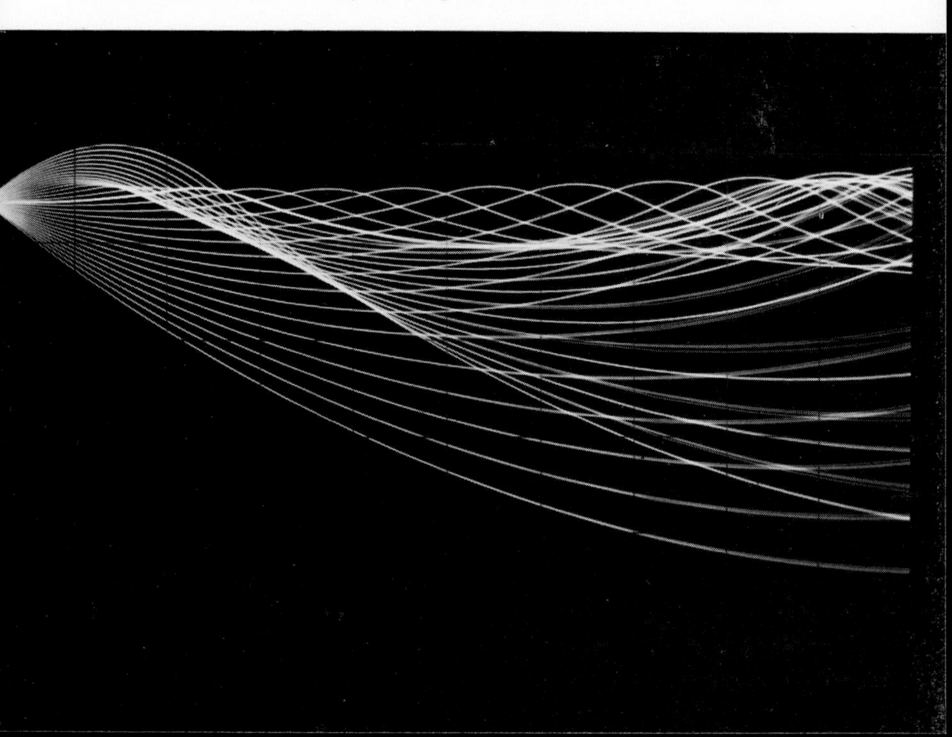

Page 52 *Development in echosounder recorders.* (*above*) The Admiralty Research Laboratory instrument used for trials at Sheerness in 1929. (*below*) A typical fishing echosounder in current production

THE POSTWAR YEARS

For security reasons it is impossible to carry this account very far into the postwar period. The U-boat threat in World War II could never have been contained without sonar. On the other hand the limited underwater endurance and speed of the submarines of those days forced them to move on the surface to take up favourable attacking positions; and the development of the 'schnorkel' though it reduced, did not eliminate, the risk of surface detection when recharging batteries. As a result, many U-boats were sunk after being picked up on the surface by shipborne and airborne radar.

Nuclear propulsion and air recycling systems for life support have turned yesterday's submersible into today's true submarine. In future, sonar detection systems will be unaided by radar, and in other respects, too, the balance of advantage has moved towards the submarine. Submerged speeds of 6 knots maximum, and then only for short periods, have been increased to 30 knots or more (the exact figures are not published) for an almost indefinite period. This means that in normal sea conditions in the North Atlantic a nuclear submarine is uncatchable by any surface ship. Great strides, too, have been made since World War II in the performance of torpedoes.

All this may be counterbalanced to some extent by improved sonar detection ranges, but here the physics of sound in water set fairly definite limits upon what is ultimately possible. Variable depth sonars (VDS) can greatly improve performance in adverse conditions. In the presence of a negative temperature gradient starting at the surface, such as may occur in calm sunny weather, the beam is bent downwards and detection of targets within 200–300m of the surface will be very poor with a shipborne transducer. The improvement obtained by lowering the transducer is demonstrated in the photographs of a ray path analyser display shown in the plate on p 51. In suitable conditions ranges may also be increased by reflection from the seabed with 'bottom bouncing' sonars, but the successful use of this technique as well as VDS implies detailed and up-to-date knowledge of sound velocity gradients, seabed contours and reflectivity.

New types of anti-submarine weapons, such as 'Ikara', which consists of an airborne missile designed to enter the water in the vicinity of the target and then home on to it by acoustic methods, have a greatly increased range performance. But even when armed with such sophisticated systems as this, anti-submarine vessels in the numbers available in World War II would still be insufficient to provide the same degree of convoy protection against the high-speed nuclear submarines of today.

In some naval circles the view has long been held that the only way to overcome this disparity is to set a thief to catch a thief. The 'hunter-killer' type of submarine is the outcome of this line of thought. From this point of departure it is possible to imagine some time in the future a form of totally submerged naval warfare in which there would be as many different classes of submarine as there are surface warships today. One may pause to wonder how many nuclear-powered submarines are, even now, stealthily following each other about in the dark confines of the oceans; but such thoughts, doubtless fascinating to the naval strategist, cannot usefully be pursued in public debate. The systems that may become available are—and are likely to remain—as opaque to the layman as the medium in which they would operate. It is nonetheless a chilling concept.

Helicopters, despite their limited endurance, offer another method of protecting convoys and task forces against the threat of submarine attack. Fitted with a 'dunking' sonar they can hover while lowering the transducer into the water for an all-round search, and then move forward to repeat the search further on, adopting the procedure used with hydrophones in World War I. High-speed scanning systems, which are described in Chapter 4, are well suited in principle for this sort of work.

It can be argued, of course, that in a future world war Armageddon would be reached before a single convoy had time to cross the Atlantic and so its protection would hardly matter. It may be, however, that the button would not be pressed, in which case, as things stand today, the convoy would have a hard time of it.

Notes to this chapter are on page 192

3

NAVIGATION

THE ECHOSOUNDER

UNDER THE HEADING OF NAVIGATION ONE THINKS AT ONCE OF THE echosounder in the acoustic context, and this instrument will provide the theme for this chapter, as did sonar for the previous one. Echosounders began to replace sounding machines in both merchant ships and naval vessels in the 1920s, preceding radar and radio position fixing systems by a quarter of a century, and so they were relatively more important as a navigational aid before World War II than they are today.

The great advantage of the echosounder over previous methods of finding the depth lies in its ability to provide a continuous record of depth at a rate limited, in principle, only by the speed of sound and the depth of water. The information may be displayed in various ways, the most common being on a chart which shows not only the present depth as a mark against a vertical scale but also all the previous recordings over a period of time. This continuous profile of the past history is particularly useful when making a landfall. The rate at which the depth is decreasing can be compared with the navigational chart and outstanding features on the seabed can easily be recognised. An example of such a feature is the fault that runs diagonally across the western entrance to the English Channel known as the Hurd Deep, and an early record of this will be found in the plate on p 85. The present depth may also be indicated by means of a pointer, or digitally, or the system may be used to operate an alarm at a preset depth.

EARLY DEVELOPMENTS

During 1919–20 a project to develop a diaphragm sounder was undertaken. The sound pulse was derived from a laminated hammer acting as a solenoid. With the current switched on it compressed a spring. When the current was broken, the hammer was driven violently into contact with the central boss of a 10in (25cm) diaphragm, causing it to vibrate in a complex way and emit a broad-band pulse centred at about 500Hz.

This principle was used in what was known as the *shallow water echosounder*, developed by the Admiralty Research Laboratory (ARL), at Teddington, from 1921 onwards. In this echosounder the diaphragm was mounted in a tank inside the ship's hull filled with fresh water, and the receiving hydrophone was placed in a similar tank on the opposite side of the keel. The hammer was designed to strike three times a second by means of a commutator with an insulated segment rotated by a constant-speed motor. The most memorable feature of the design was the provision of a second commutator, rotating at the same speed, which served to short-circuit the telephones in the receiving circuit except for a short period during each revolution. By this means the sounds recorded by the hydrophone could only be heard at full strength momentarily between successive blows of the hammer. The telephone pick-up brushes bearing on the commutator could be displaced by hand until the echo from the bottom could be heard at maximum volume. The required displacement measured the elapsed time between the transmission and the return of the echo, giving the depth, which was read off on a dial. The effective range of the set was limited to 150–200 fathoms, and depended critically on the siting of the transducer tanks. Quantity production began in 1925 and a number of naval and mercantile ships were subsequently fitted.

These sets worked quite well, but failures of the spring and hammer were predictably frequent. In a deep-sea version of this design, which was soon introduced, the diaphragm was replaced by a resonant steel rod which was struck, as before, by an electromagnetically (later pneumatically) operated hammer. The lower end in water produced a pulse consisting of a damped train of HF waves. Acoustic cavitation, experienced initially, was overcome by enclosing the rod in an oil-filled dome under pressure, dissolved air and bubbles having been removed. A tuned high frequency microphone of the double diaphragm type enclosing carbon pellets or Rochelle salt was used to detect the echoes.

During these years parallel developments were taking place elsewhere, notably in Germany, France and America. In the earliest designs the main difficulty lay not in obtaining an echo but in measuring the time interval between transmission and reception. Dr Alexander Behm, who began his experiments as early as 1911, attempted to avoid the problem altogether by measuring depth as a function of intensity, but this idea was doomed to failure if only for the fact that the reflectivity of the bottom itself varies widely. In his first successful design, which appeared shortly after World War I, the sound pulse was initiated by a rifle fired into the water. The direct signal was recorded in a hydrophone on the same side of the ship and the reflected signal from the bottom was picked up by a hydrophone on the opposite side. The former operated the starting switch for a constant-speed motor and the latter a dynamic breaking circuit which stopped it. The motor rotated a mirror through suitable gearing which reflected a beam of light on to a depth scale on the bridge.

In April 1922 the first continuous series of soundings across the Mediterranean was run from Marseilles to Phillipville with the aid of an instrument designed by the French hydrographic engineer M. P. Marti, and in this case also the pulse was made by firing a bullet into the water. For depths over 1,000m a charge was detonated in the water. This was followed by the Marti continuous sounding recorder, used in conjunction with Langevin's piezoelectric sandwich transducer. The echo was recorded by a rotating arm that scribed a line on a smoked paper. The bottom echo operated a relay which rocked the scribe so that it made a nick in the otherwise smooth line. The paper was blackened by a petroleum lamp with a smoking wick, which was moved backwards and forwards under the paper by means of a cranked lever. This unlikely apparatus really worked (see plate, p 85).

In America Dr Hayes of the Naval Research Laboratory (NRC) overcame the difficulty of measuring short time intervals by taking advantage of the very sensitive binaural faculty of the human ear. The aural sense of direction depends on the difference of arrival time of the sound at each ear, and the accuracy of discrimination is said to be approximately $0{\cdot}5 \times 10^{-5}$sec. One side of the headphones was connected to the projector and the other to the hydrophone; and the prf was then adjusted until transmission and echo coincided exactly.

Another system, called the 'angle of reflection' method, involved two hydrophones relatively close to each other near the bow and a sound source near the stern. It will be seen from Fig 8 that the bottom echo will reach hydrophone A shortly before B and this time difference is inversely proportional to the depth. Each hydrophone is connected to one side of a pair of headphones, one of them via an adjustable delay line, or compensator. By altering the delay until the two signals are exactly matched, the depth can be determined. Reliable depths were obtained by this means in water up to about three times as deep as the length of the ship, and it will be seen that the method has the advantage that accuracy *increases* as the depth *decreases*.

The author of an article written in 1925 reviewing some of these strange systems ended by stating: 'Whether this development (echosounding) is tantamount to safer navigation may remain an open question for it is very likely that navigators become more audacious as the means of navigation become more developed.' This conjecture has indeed been proved by the events of the last 50 years.

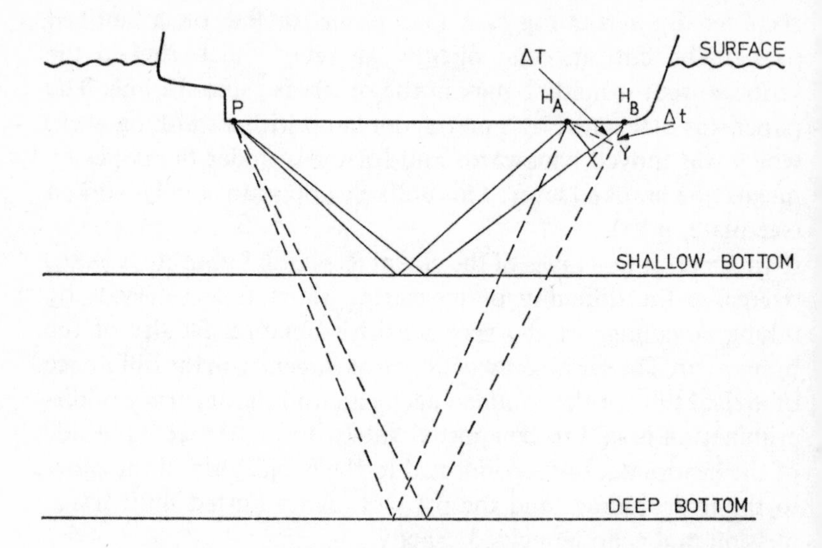

FIG 8 In this early method of echosounding a transmission is made at P and is picked up via the bottom in the two hydrophones at H_A and H_B. The time difference between the two sound paths, Δ T for the shallow bottom and Δ t for the deep bottom, is a function of the depth of water. It is measured by applying one hydrophone to each ear, the nearer via a compensator which is adjusted until the two sounds are binaurally matched, that is to say, exactly in phase. Since the direct distance between the projector and the hydrophone is known, the depth can be calculated from the time difference shown on the compensator. The shallower the depth, the more accurate the result. Reasonably accurate soundings were achieved by this method to a maximum depth of three times the length of the ship

The major advance in echosounder technology came with the adaptation of the magnetostrictive effect. It seems to have been largely a matter of chance that this method was developed in Britain for echosounding in preference to the piezoelectric system that had already been developed for sonar and, in France, for echosounding. Sonar work in Britain was highly classified and was being undertaken at a different establishment.

There certainly appears to have been no comparative technical evaluation made at that time, and today both methods are used in various applications for both echosounding and echoranging. The basic research and development work was carried out at ARL in Dr Wood's underwater acoustic department.

Dr Wood, about this time, also designed the 'contactor' type of transmitter, in which a large capacity condenser is charged to a suitably high voltage between transmissions, and the pulse is achieved by releasing a spring-loaded contact to discharge the condenser directly across the low impedance nickel pack winding. The result is a heavily damped train of oscillations at the resonant frequency of the transducer, leading to a powerful pulse of sound of about 1ms in duration.

Also at ARL the first successful chemical recorder was developed to give a continuous record of bottom echoes. In this system an electric motor is used through suitable gearing to draw a pen at uniform speed across a roll of paper chemically treated to discolour when a current is passed through it. The pen is electrically connected to the output of the receiving amplifier, the circuit being completed to earth by the support or 'platten' on the other side of the paper. The sound pulse is transmitted as the pen crosses the leading edge of the paper. Assuming a fixed value for the speed of sound, accurate enough for navigational purposes, the point reached by the pen when the bottom echo returns to the transducer and marks the paper is a measure of the depth. This can be read directly from a scale placed across the width of the chart. The roll of paper is pulled slowly at right-angles to the pen's travel so that successive echoes are laid alongside each other, and in this way a continuous profile of the seabed is produced. The chemical paper is soaked in a solution of potassium iodide and starch, as used originally by Faraday over 100 years ago. In the first designs the paper was moistened by capillary attraction from a wick, but later the potassium iodide paper was supplied moist in air-tight containers and similarly held in the recorder pending use.

All these new features were brought together in the first British recording ultrasonic echosounder, which had its sea trials at Sheerness in 1929. It drew a profile of the seabed across the dredged channel in the Thames estuary which is typical of, and hardly distinguishable from, the records made with today's machines. Toroidally wound packs of ring stampings were used for transmission and reception in inboard tanks, one either side of the keel, with air-filled conical reflectors (see plate, p 34).

The recorder was originally designed to cover the depth range 0–150ft. For further trials in deeper water the recording speed was reduced to one-sixth to convert the scale from feet to fathoms, and provision was made for advancing the transmission in relation to the pen to enable the chart record to be extended in two further steps to show 150–300 fathoms and 300–450 fathoms. This technique, sometimes referred to as 'phasing', takes advantage of the fact that the pen is off the paper for the greater part of the interval between transmissions.

This instrument represented a major technical breakthrough which was to have immense commercial potential—the high-water mark of echosounder development, in fact, and the starting point for a long series of models manufactured by Kelvin Hughes. While these have incorporated many improvements over the years, the essentials of the design have remained unaltered (see plate, p 52).

METHODS OF DISPLAY

For general navigational purposes the echosounder is usually more valuable as a warning device than as an aid to position fixing. The method in which the depth information is displayed is, therefore, of special importance. Its application to the problem of navigating very large bulk carriers in channels and straits with the minimum safe underkeel clearances is somewhat specialised and will be more conveniently dealt with separately. This applies also to its use in deep water outside the 100 fathom or 200m line, which is of less interest for general navigation.

The chart recorder, already described in principle, has the outstanding merit of displaying past as well as present depths—of possessing a 'memory'. The alignment of successive traces one alongside the other is also important because this trace-to-trace correlation causes repeated echoes, such as are returned from the seabed, to stand out against a random noise or reverberation background even when the amplitude of the wanted echo is no greater than the ambient level. In other respects the chart contains more information than is apparent at first sight, and more than can be displayed in any other way. A trench filled with mud in precipitation can be recognised immediately for what it is, and echoes from objects other than the seabed itself in the way of the beam, such as mid-water fish shoals, and from aeration, can easily be identified. This subject is pursued more fully in Chapter 4, where it will be shown that the art of chart interpretation has an extensive repertoire.

The method of 'phasing' the transmission already referred to is widely incorporated in present-day echosounders. In glancing at the chart the Master may assume that the part of the water column shown represents depths between, say, 20 and 45m, not realising that this has been altered to 0–25m to keep the trace on the chart. He would then assume the depth to be 25m more than it really is. Ships have been put at risk in this way before now.

Or it may happen that the second echo, caused by the sound pulse being reflected via the surface to the bottom a second time, and therefore marking the chart at twice the true depth, will be mistaken for the first. This, of course, can only occur when the true depth is less than the minimum shown on the scale, so that the second echo is the only one visible. In shallow water the second echo can produce a saturation mark on the paper, but usually appears as a double line, a part of the pulse having travelled the slightly shorter distance via the hull of the ship. No experienced navigator ought to confuse one sort of echo with the other. Nevertheless it can happen.

None of the several types of recording paper now in use are entirely satisfactory. Potassium iodide paper has a limited shelf life, particularly in the tropics, and must be kept moist before use. It has the largest dynamic range as defined by the number of 'just noticeable differences' (JND) between no mark at all and saturation. Dry papers, such as 'timefax' and 'teledeltos', which have a carbon base are somewhat less sensitive and have an unpleasant smell, but have the advantage of making it possible to mount the instrument with the pen running downwards across the chart so that vertical depth is shown against a vertical scale. This, the logical arrangement, does not work with moist paper, because it then dries unevenly and is liable to tear.

There are advantages in having a repeater on the bridge, set in a prominent position, and putting the recorder in the chart room. However, it is technically difficult to design a repeater, which may present the information in digital or analogue form, to be completely reliable. Time and amplitude gating circuits are used to select and lock on to the bottom echo, and these have to be designed in such a way as to ensure that an echo from a fish shoal directly beneath the ship, which may well be of greater amplitude than the bottom itself, is not selected in error. Especially difficult cases can occur on encountering a trench that may be filled with mud in precipitation or other sediment with a high porosity, since an echo will be reflected from the layer of mud but enough of the sound will penetrate to the hard bottom below to give an additional echo. The echo chart shows the picture, but the digital read-out can only follow one echo at a time and cannot even indicate the existence of a second one (see plate, p 104).

The bridge unit should also incorporate a 'fail-safe' device so that as soon as the bottom echo is lost for more than a few seconds, due to excessive aeration round the transducer while going astern or in exceptionally bad weather, the read-out displays a sign which says in effect 'Disregard me' instead of simply remaining at the last definite reading or reverting to zero. A rather crude alternative, which bypasses these problems at the expense of precision, is the 'flashing light' display: this consists of a rapidly rotating disk synchronised to the transmission, with a neon light, mounted radially, which flashes on receipt of an echo. The depth scale is marked round the circumference.

Designs also exist of an alarm unit that can be set to give a warning when some selected depth of water is reached on approaching land. This, too, is not without its limitations even today.

UNDERKEEL CLEARANCE

If, in general, it is true today that the echosounder is of somewhat secondary importance as a navigational aid, this is not so for supertankers steaming in shallow waters, such as the Straits of Dover, with an underkeel clearance (UKC) of only a metre or two. The echosounder then becomes a key instrument and special systems incorporating four transducers—one forward, one aft and one on each beam amidships—have been designed and installed. The great size of these hulls makes this necessary. The depth profile for each recorder is shown by a thin line and the four signals can be superimposed on one display. This means the effect of rolling and pitching on the overall draught can be precisely measured, which is important since a roll of 3°, scarcely noticeable onboard, would increase the draught by nearly 4½ft (1·3m) at the outboard edge of the flat bottom.

In restricted and shallow waterways the UKC is reduced and the trim altered in any large deep-draught vessel by the effect known as 'squat'. Whenever the space beneath and on either side of the ship is limited, a reverse current is set up by the water escaping past the ship as she piles the sea up ahead in moving forward. The more restricted the space and the faster the ship goes, the more marked is the effect.

Bernoulli's law states that in a fluid pressure decreases with increasing velocity, and so this reverse flow under the ship reduces the pressure and the ship sits down or 'squats' in the water. The effect varies with the shape of the hull and the configuration, including width and depth of the channel, so that it differs for every class of ship and every restricted waterway. It is further complicated by the fact that a certain amount of trim by the head is also introduced. It will be evident that the effect reduces the speed. Conversely, more power is required to maintain the same speed. Manoeuvrability is, if anything, improved. It is in fact relatively difficult to run aground on a soft muddy bottom under these conditions, though where rocks are present the situation is very different. Therefore, in deciding how much allowance should be made for squat, the nature of the seabed must also be taken into account.

Squat values worked out from the extrapolation of results of model experiments have since been shown to have been over-estimated by an extended series of trials in the early 1960s in the Maracaibo Channel at the head of the Gulf of Venezuela. For these trials an echosounder transducer was fixed to a platform on the seabed facing upwards, so that the local depth of water could be compared accurately with the UKC of each ship as she passed overhead. From this data the amount of squat and trim by the bow was derived for a range of speeds and different ships. A 200,000 ton tanker's earning capacity is said to be of the order of £25,000 per ft of draught, so that accuracy is worth a lot of money.

The allowance for squat to be made when deciding a maximum safe draught to which the ship can be loaded no longer presents problems when the channel or shallow passage to be negotiated is in still water; but when the effects of squat have to be added to those of the sea and swell in increasing the draught, the problem is less simple. Attempts have been made to correlate swell states with draught increase, but without full success so far. The number of variables is extremely large and the relationship between wave patterns and ship movement is complex. Many hundreds of hours of tanker time would be needed for practical experiment before anything approaching a worthwhile analysis could be made. Fortunately, however, the worst wave conditions that will cause no significant movement for each class of vessel can be demonstrated with existing instruments. Practical tests have also indicated that a given UKC can be established as satisfactory up to a known sea condition specified in terms of maximum wave height. In all this work the echosounder, specially designed to read and display very near echoes, plays a central role both as a research tool and as a monitoring instrument for ships in service.

A low minimum depth presents no great problem. Dr Wood's shallow water echosounder achieved a minimum depth of 1ft, and this can be improved upon by the use of a higher frequency. Since a general purpose navigational echosounder would normally be carried also, the multiple equipment for measuring UKCs would require no great depth capability and could well have a frequency of 200kHz or even higher. A very short pulse is essential, implying a wide frequency band and a consequent preference for a relatively non-resonant (low 'Q') transducer. Separate transducers for transmission and reception are required.

DEEP WATER

Turning from the very shallow to the very deep in echosounder performance, it is certainly true that a profile of the seabed outside the limits of the continental shelf is of little interest to the navigator at present. It does not help him to fix the ship. However, the bottom of the ocean is not so featureless as was once supposed, and there are areas which are as mountainous and fissured as anything to be found on land. It is not too fanciful to look forward to the day when the mapping of the seabed floor is more advanced and it will be possible to produce ocean charts containing sufficient detail for accurate navigation, at any rate on certain routes. In the meantime echosounders have been developed with all the refinements and special features needed to return echoes from the deepest places in the sea. These are fitted in all ocean-going survey ships, oceanographic research vessels and cable-laying ships.

Just 21 years ago the Danish Research Vessel *Galatea* obtained a record of soundings at 10,540m in the Philippine Trench, the greatest known depth in the sea. Since the speed of sound in water is 1,500m/s, one echo would be received every fourteen seconds, in which time a ship at 15 knots will have covered 100m. Furthermore the transmitted beam of the typical echosounder is only partially directional, and after travelling downwards through the water for so great a distance it becomes dispersed over a wide area. Thus the standard echosounder with sufficient power to return echoes from the greatest known depths in the sea will still be unable to draw anything approaching a profile of the sea floor, because the information rate is much too slow and the bearing discrimination insufficient to show any fault or trench or other discontinuity on the seabed not much smaller than the Grand Canyon. To get these depths a relatively low frequency is required, and this aggravates the problem still further because the transducer has to be large.[1]

In such deep water even a relatively narrow beam would have to be stabilised against roll and pitch to be certain of getting continuous echoes in rough weather. The transducer can either be mounted on a stable platform or 'steered' electronically.[2]

Some improvement to the very low information rate can be had by a multiple transmission system in which several pulses of sound are travelling through the water at the same time, but this technique brings with it its own problems. For instance, the bottom echo may be masked by reverberations of greater amplitude being received at the same time from a more recent transmission.

A number of designs of precision depth recorders have been developed in recent years for deep ocean sounding.[3] They are collectively called PDRs, for having a more precisely controlled prf than is usual for a standard navigational machine. A crystal or tuning-fork is normally used for this purpose to reach an accuracy of 1 part in 5,000.

In the majority of these recorders the moving pen is replaced by the moving point of intersection between a helix and a straight bar. In this way the 'dead time' can be eliminated entirely, the whole period between transmissions being shown on the chart. When used in a multiple transmission mode, the scale does not directly indicate the depth of water, as each bottom echo is derived from some transmission pulse other than the immediately preceding one. Special provision must therefore be made to identify the multiple of the full scale that must be added to the indicated depth to arrive at the true figure. One way of doing this is to revert briefly to the single transmission mode at regular intervals, when the true depth can be read approximately on a much compressed scale. The optimum multiple transmission programme, ie, the one that yields the highest overall prf without obscuring the bottom echo trace, depends on the depth. It can be adjusted manually each time the trace passes through 'zero', or automatically by means of an electronic lock-on circuit, provided that the signal-to-noise ratio is good enough.

INTERGOVERNMENTAL MARITIME CONSULTATIVE ORGANISATION (IMCO)

A curious event in the recent history of echosounding is the decision by the IMCO assembly to add a recommendation to the International Convention for the Safety of Life at Sea (1960) stating that, 'All new ships of 500 tons gross tonnage and upwards, when engaged on international voyages, shall be fitted with an echosounding device.' Coming some 40 years after the general introduction of this instrument to merchant ships of all flags it seems a little late in the day. Moreover, it will be several years yet before a specification can be agreed by all the members of IMCO and the proposal ratified by a sufficient number of countries to make it mandatory. It is impossible to say, of course, how many ships would have run aground if they had not had echosounders to warn them of the danger. What does seem clear, however, is that the number of groundings in proportion to the number of ships at sea has remained fairly constant over the years. Furthermore an analysis of those that have been investigated, probably only a small proportion of the total, indicates that the echosounder has not had much influence on the outcome. It was either ignored, sited where it could not easily be seen, or else the ship struck a pinnacle of rock when the echosounder was indicating, quite correctly, a depth of several metres. There have been no recent reports, if any at all, in which the cause of grounding has been attributed to the lack of an echosounder. There have been cases, on the other hand, in which the misreading of the recorder has, at the least, been a contributory factor.

The Department of Trade and Industry (DTI) in Britain has now put forward its recommendations for a set of minimum performance standards for a navigational echosounder, following the IMCO resolution, and other countries will no doubt soon do likewise. The DTI specification includes environmental standards and tests which follow closely those established 20 years ago for marine radar. It also includes a 'figure of merit' method of assessing performance based on the sonar equation.[4] As to the specification itself, the accent is certainly on the word 'minimum'. It is a lowest common multiple of a specification. To quote one example, the minimum depth the instrument is required to register under normal propagation conditions is 2m, which compares unfavourably with the 1ft achieved by A. B. Wood 40 years ago. It has been a theme of this chapter that since, today more than ever, the navigational echosounder is essentially a warning device, the method of display is of special significance. On this subject, apart from making the chart recorder obligatory and referring to 'fully adequate' illumination, the specification has little to say. It does not, for example, prohibit the provision of depth scales that do not start at zero, it does not specify the minimum distance at which the scale must be legible and it does not say where the recorder is to be fitted.

One may wonder if such a mouse of a specification will ever save a single ship.

DOPPLER ACOUSTIC LOG

This type of log has been developed for commercial use only within the last few years, and is not yet widely fitted. It is unique in measuring speed over the ground rather than speed through the water. The doppler effect is described in note 5, Chapter 2 (p 193). In this application directional sound pulses are transmitted obliquely ahead of the ship towards the bottom. The frequency of the bottom echo returned to the ship is raised by a factor equal to $\frac{2V\cos\theta}{C}$ to a first approximation where θ is the angle of depression of the beam, normally 60°, V is the speed of the ship and C the speed of sound in water. The width of the beam in the vertical plane, though narrow, is still wide enough to return an extended echo and a varying frequency shift from which the spot frequency required for the computation has to be derived.

Changes in the frequency shift are introduced by a sloping bottom, by pitching and by refraction of the beam. All these sources of error can be reduced, if not eliminated, by transmitting the pulse also in the after direction with the beam tilted at the same angle of depression to give a frequency shift of the opposite sign. The frequencies of the two returning echoes are matched to give a doubled effect. This is known as the Janus configuration after the Greek god who faced both ways. In this way the speed of the ship over the ground is directly measured and by integration the distance run is obtained.

The system is limited by the greatest depth at which a bottom echo can be obtained. This is at present claimed by the manufacturers to be of the order of 600 to 1,000ft (180 to 300m) so that the log can give ground speed on the continental shelf but not over the abyssal plains of the deep ocean.

In most systems the transducers are tilted at the appropriate angle. A German Atlas design uses a horizontal transducer array obtaining the necessary angle of transmission by means of a 3-phase supply which is used to produce a phase shift across the successive elements that form the transducer. The precise angle at which the beam enters the water depends on the speed of sound, C. As this is also a factor in the doppler shift equation C cancels out and therefore does not have to be measured and allowed for as in fixed angle systems. The flat faced transducer has another advantage of reducing flow noise across the transducer face when the ship is proceeding at more than a few knots.

Both pulsed and continuous wave systems have been developed. The former can be used at greater depths and the latter has some advantages for use in very shallow water. When the amplitude of the bottom echoes falls below a certain level, the log measures the frequency shift of volume reverberations and thus gives speed through the water in common with all other kinds of log. Another development involves locking on to the deep scattering layer, as an alternative to the bottom, in deep water.

A major advance in the technique would be the ability to echo from the bottom in all ocean depths. This is not possible at present because the requirement for a very narrow beam as free as possible of sidelobes conflicts with the need for a low frequency for long-range performance. Non-linear acoustics may possibly provide the solution. The Raytheon FADS echosounder referred to in Chapter 5, which uses this technique, could be adapted for this purpose. The system is designed to produce a 2° beam without sidelobes at 12kHz, which would represent a major navigational breakthrough, as it could provide a system of dead reckoning significantly more accurate than anything in existence at the present time, except for the Ship's Inertial

Navigational Systems (SINS) which is, however, too costly for mercantile use. In the meantime the doppler log is being widely fitted in oceanographic research vessels where position fixing is of special importance, interfaced with satellite navigation or the Omega vlf hyperbolic system.

These logs also have some naval applications, and are used for dredging and in some deep submersibles. The doppler log has also been tried but with limited success as a method of measuring the approach rates of very large tankers when berthing alongside a jetty. For this application the fore-and-aft sound beams are backed up by additional pairs looking athwartships, one mounted forward and one aft, and a combination of these readings gives the sideways motion of the vessel and the rate of swing. An approach velocity of a few feet per minute by a supertanker is sufficient to cause damage, and so, to be of value, the logs must be very accurate and capable of registering small velocities. Typically, they have digital read-outs scaled in knots, tens of feet per minute and even feet per minute, but one wonders if they are genuinely capable of such a low threshold. Tug-generated turbulence can upset the readings during the final stages of the approach, at the very moment in fact when they are most needed. Doppler radar provides an alternative solution to this particular problem, and is probably the more effective.

Doppler logs are at present manufactured by EDO Western, Marquardt and Sperry in America, by Atlas Elektronik in Germany and IFP in France.

UNDERWATER ACOUSTIC POSITIONING

In the early days of echosounding it was thought, at any rate by one or two enthusiasts such as Arthur Hughes, the friend of navigators, that the more detailed maps of the sea floor would open the possibility of using the echosounder profile for geographical location. But there are few natural signposts in the seabed as conspicuous as the many landmarks onshore, ie, when 'viewed' by an echosounder.

Although this particular dream has proved to be an illusion, the science of underwater acoustics in a different form is being increasingly applied to the problem of position fixing, particularly in the deep oceans where no other means exist. The vlf electromagnetic waves used by the worldwide hyperbolic system known as 'Omega' penetrate a few feet into the sea and can therefore be used by submarines near the surface. This may be of some military consequence, but from the point of view of oceanography and geodetics manned or unmanned submersibles and other instrument packages near the surface can best be positioned by relation to a mother ship, in which a satellite system would normally be used (when operating beyond the scope of accurate near shore location devices such as Hi-fix) interfaced with a Ship's Inertial Navigational Systems (SINS) or a doppler acoustic log. In deep water position cannot be directly determined by relation to a surface ship even in the case of a 'towed fish' or a submersible connected by an umbilical cord. What is required is an acoustic tracking system of some sort that can relate the survey or mapping instruments to reference points on the sea floor. Acoustic signals have been used for navigation and positioning at sea to a limited extent for many years, but it is only within the last decade that activity in the development of underwater positioning systems has reached significant proportions. A growing interest in the deep oceans and in the seabed beyond the continental shelf areas has led to a number of studies and much practical work in this new field of endeavour.

The first step in discovery is to make a map, and today we are at the same threshold on the ocean bed as were the great explorers and mapmakers such as Cook and Flinders in the Pacific 200 years ago. It is perhaps unfortunate that by its very nature this work of exploration is relatively inaccessible to the man in the street. It is none the less significant. So far much the greater part of it is being carried out by American organisations such as the Scripps Institution of Oceanography and the Woods Hole Oceanographic Institution. This is partly a question of resources and partly through impetus given by the tragic loss of the US nuclear submarine *Thresher* in April 1963 during a test dive after a refit, in 8,400ft of water. The event demonstrated as nothing else could have done the dearth of submersibles and instrument systems capable of working at these depths, still relatively moderate when compared with the oceans as a whole. The emphasis in the Deep Submergence Systems project that arose from this traumatic experience has quite naturally been on the development of submersibles capable of working in such depths. Less has been heard of the equally vital technology of erecting the signposts on the sea floor without which any submarine craft is completely blind.

As with electromagnetic systems, the most obvious distinction between the various methods of position fixing that can be used underwater separates those which measure the travel time between source and receiver from those which measure differences, either of the time of arrival of the signal at the receivers, or of two synchronised signals at a single receiver. In the former group the position lies on the surface of a sphere and in the latter on a hyperbola of revolution. Three or more such measurements in each case serve to fix the submersible. It will be seen that this is essentially a three-dimensional problem, since the object to be fixed may be anywhere between the surface and the seabed. However, the vertical co-ordinate can be measured directly by an upward looking echosounder or pressure gauge. The former can give an accuracy of a metre or better, but there is a lack of suitable pressure gauges with adequate resolution at great depth. Then the position is fixed at the intersections of the three-dimensional shapes, with the horizontal plane surface at the depth indicated, which is the same as the surface problem. The receiving hydrophone may be fixed to the sea floor, though not necessarily resting on it, or it may be mounted in the mother ship or on the submersible itself. The first of these alternatives is generally more appropriate in shallow water, where a number of different vehicles may be using the system at the same time. These three choices, in conjunction with the hyperbolic 'range bearing' or the 'two range' configurations, or with various combinations of them, add up to a total of thirty possible systems combinations. Only about a quarter of these have been used so far in the open sea, although a number of others are under development. The choice of the most suitable system to meet any particular requirement is, therefore, still a very open one, with relatively little previous experience to go on.

HYPERBOLIC SYSTEMS

Much the earliest purely acoustic system falls into the hyperbolic category. This is the Sound Fixing and Ranging (SOFAR) system used during and after World War II for locating the positions of ditched aircraft. It takes advantage of the exceptional sound propagation conditions that are found in the deep sound channel. The sound from a small charge detonated near the axis of the channel suffers no attenuation through scattering, and in consequence can be detected at distances well in excess of 1,000km by hydrophones placed at the same depth. These are connected by cable to shore stations at which the recorders are activated by the leading edge of the signal. The geometry of the sound channel is such that the sound ray travelling directly along the centre of the path has the slowest velocity and is the last to arrive, although going the shortest distance. The characteristically sudden cut-off following the climax of the slow crescendo makes it possible to measure the time to an accuracy of $\frac{1}{10}$sec. The signal is recorded at three or more widely separated hydrophones, each with an accurate time signal. Time differences are measured between pairs of hydrophones to fix the sound source. In this particular case the depth is not a variable in the equation, and the system is thus exactly analogous to a surface hyperbolic system. The chief limitation to the overall accuracy is an exact knowledge of the mean velocity of propagation. In common with all methods of fixing, the angle of cut between curves is also significant. Total errors can normally be kept within 100m. SOFAR installations have been set up in both the Pacific and Atlantic Oceans. The system has also been used for fixing the locations of submarine seismic disturbances.

The other pure time difference system is the US Missile Impact Location System (MILS) used by Atlantic and Pacific missile ranges. For this purpose groups of six hydrophones are disposed over a 10 mile area and the cables brought to a common point. Time differences in arrival of the 'splash' signal are computed to fix the splash point of the missile to within 10m or so, provided that it falls within the ring of hydrophones. Beyond these limits the accuracy of fix is reduced.

RANGE-BEARING SYSTEMS

Systems in this category incorporate either a directional scanning receiver or pairs of hydrophones mounted in a three-dimensional array. The latter arrangement has been used in a successful experiment in Puget Sound in which the transmitting element together with an array of four hydrophones arranged in three orthogonal pairs was mounted on the seabed and cable-connected to the shore. The transponder was mounted in the submersible and the range and bearing from the seabed rig was plotted automatically. In shallow water (200m) this method yielded accuracies of the order of 1m. When used in deep water at greater range, it was necessary for the acoustic operating frequency to be reduced, so that the accuracy was less. The same principle has been used with the transmitter and hydrophone array mounted on the mother ship. In one such configuration the mutual separation between hydrophones was increased by hanging the third one from the end of a boom 15–20m off the centre line of the ship. The disadvantages of this system are in the poor signal-to-noise ratio, especially in bad weather; in the errors caused by the movement of the ship in a seaway; and, of course, in the necessity to resort to a secondary, surface, system to fix the position of the ship itself. This was the approach adopted in the search for *Thresher*. The surface ship acting as an intermediary navigational base, itself fixed by Decca and Loran C, was used to establish positions of transponders planted on the sea floor, and of towed instrument packages and submersibles such as the bathyscaphe *Trieste*. This kind of system has the advantage of mobility during searches but is dependent on the accuracy of the surface system in whatever area is involved. In the search for *Thresher*, though she went down close to the United States, detailed work was interrupted at night because of the unreliability of electromagnetic propagation in the locality. These difficulties emphasised the need for bench marks on the sea floor for accurate and consistent survey work.

RANGE-RANGE SYSTEMS

The most straightforward method of tracking an underwater vehicle in deep water is in relation to a network of transponders or precision oscillators anchored to the sea floor. Provided that the signal-to-noise ratio is adequate and that due attention is paid to the effects of refraction, the arrival time of pulses can be determined to a millisecond, corresponding to an error of between 1 and 2m. Transponders may be expendable or recoverable. Battery-operated types have a life measured in years and millions of responses, and atomic energy power sources open the possibility of virtually permanent operation. Transponders are interrogated in sequence, each responding to a particular frequency or code. The transmitted interrogation pulse would normally be selected automatically by a programmer. Alternatively, all the transponders in a particular group can be activated simultaneously, each replying in a different frequency. The received signals can be displayed on a recorder from which the travel times for individual signals from each transponder can be read. The advantage of doing this instead of applying the intervals directly to a computer is that it permits visual averaging, with a consequent improvement in accuracy, especially in poor acoustic conditions.

Fundamental to the method is an accurate determination of the positions of the transponders themselves. When operating in deep water, say thousands of metres, it is not possible in practice to lay them in precisely determined positions. During the time taken by the transponder assembly to reach the seabed, which may be the better part of an hour, it can drift a long way. After a preliminary survey of the area by echosounder to select a suitable site, the first transponder is dropped, followed by the remainder after steaming the distances and courses required to achieve the approximate configurations and density already decided upon. The precise positions of the transponders are determined subsequently by triangulation methods analogous to those used ashore for establishing beach marks. This is a complicated process involving hundreds of simultaneous readings best carried out with the aid of a computer, the purpose of which is to find the best fitting calculated positions. The self-consistent pattern of transponders can then be tied precisely to a geographical reference by a series of runs in which the ship is navigated simultaneously in relation both to the transponder pattern and to one or more satellite or electromagnetic position-fixing systems. Surveys of this nature undertaken by the Marine Physical Laboratory (MPL) of the Scripps Institution of Oceanography have shown that a mean position error of 5m radius can be achieved in depths of up to about 4,000m.

The future will see a wide variety of underwater acoustic systems, and a demand for the necessary hardware is certain to grow. Gaps in the range of instruments available at present include an inexpensive pinger with an accurate time base having a drift of, say, not more than 1 part in 10^8, equivalent to 1m/sec per day. Finally, it is the acoustic transmission problems which, indeed, lie at the heart of the matter. Successful solutions will always depend to a large extent on a thorough preliminary study of the propagation conditions in the selected area. Ray paths must be determined on the basis of measured or calculated acoustic velocity profiles, and propagation losses and ambient noise levels must be carefully studied. The sea is full of noise and intensity, and frequency peaks vary widely from place to place. The essential parameters for any new system, such as operating frequencies, separation of transponders and so on can only be settled in the light of local acoustic knowledge.

SHALLOW WATER SYSTEMS USING TRANSPONDERS

In Britain recent development of transponders and pingers has tended to follow the needs of off-shore oil exploration in the North Sea, though they will certainly have wider application in the future for navigation in swept channels and congested waters. UMEL, among other British companies, is active in designing a range of these instruments. A recent example is a light-weight model especially designed for marking wellheads which may also be attached to underwater vehicles, or even divers. It has a long operating life of 3½ years or 2 times 10^6 responses which has been achieved without a corresponding weight penalty by designing the quiescent current drain down to an absolute minimum.

The range and bearing of the marked object is obtained by a ship-borne digital sonar developed commercially by Marconi Marine from a research project undertaken by the University of Birmingham. The ranges and bearings of any transponders lying within the scope of the instrument (30° by 2km) are shown in plan view on a variable persistence storage tube. The bearings of the incoming signals are measured by means of a new principle, radically different from that used in high speed scanning systems, described in the next chapter. A method of phase correlation is used in which all amplitude data is rejected. The phase shifts between adjacent elements of the receiving transducer array of incoming signals are measured and then checked for spatial correlation across the face of the transducer, and for temporal correlation between successive echoes. Only if the preset spatial standard is met during three successive transmissions is the echo recorded. The chief advantage of this new system, called Sonafix, is that the bearing discrimination does not depend on the size of the transducer array. It is thus possible to achieve a good angular resolution without having to resort to a high frequency, and therefore short range, performance, or alternatively a very large transducer.

Notes to this chapter are on page 193

4

FISH DETECTION

IN THE GOSPEL ACCORDING TO KELVIN HUGHES THE HISTORY OF echo detection of fish began with the recordings of cod made by Oscar Sund in 1935 in the Norwegian Fisheries Research Ship *Johan Hjort*. These results were published in *Nature* in June that year. On the other hand, those who have been brought up on Marconi's Gospel will prefer to remember that it was Skipper Ronnie Balls who first saw fish echoes in his herring drifter *Violet and Rose* in 1933, and subsequently published his records showing a clear correlation between echo and catch. In fact an earlier published recording has come to light in the periodical *Stavanger Aftenblad* (8 June 1934), which shows some sprat shoals near the surface obtained by the Norwegian skipper Reinert Bokn, who proved his point by making a haul of 68 hectolitres. But there were those who preferred present prosperity to the fame of posterity and it is a fair bet that the earliest records were carefully destroyed. A well quoted story, probably not apocryphal, tells of a now anonymous skipper who, whenever he was asked about this new-fangled echosounder of his, would proclaim it useless, saying he would throw it out as soon as he got round to it. He was in the habit of going off on his own and brought in consistently good catches, but he never had any records to show.

In the literature there are several references in the 1920s to abnormal echosounder signals which were believed to have come from fish. The French navigator Rallier du Baty refers to echoes he associated with cod shoals on his way to Newfoundland in 1926. But the first prize really goes to another Frenchman, a

Professor Portier, who said in an address to the Biological Society in Paris in 1924

> Il semble bien que les bancs denses des poissons comme la sardine ou le Hareng peuvent réfléchir les ondes ultra-sonores et qu'on pourra, par ce moyen, déterminer exactement la profondeur à laquelle se tiennent les essais de poissons sur le plateau continental préalablement sondé.[1]

How right he was!

From these first tentative beginnings before World War II the detection of fish by ultrasonic means has since become standard practice in every commercial fishing industry in the world. Hundreds of thousands of echosounders have been manufactured solely for this purpose. Many innovations and improvements in design have been introduced and standards of performance have been steadily raised. In the years 1950 to 1955 the development of sonar for the horizontal detection of fish and fish shoals in mid-water was started, and has proved to be technically successful and commercially profitable. It was once thought that these instruments, on account of their relatively high cost and complexity, would be used mainly as pathfinders, and that one or two vessels only among a fleet would be fitted for searching ahead of the remainder. Such timid forecasts have been quite confounded and today one, and often two, sonars are fitted as a matter of course in all fishing vessels equipped for mid-water trawling and purse seining. The introduction of sonar for fish detection was preceded by its application to whale tracking, a different problem altogether. As everybody knows, whales are mammals not fish, but a section of this chapter will be devoted to them nevertheless.

THE FISHING ECHOSOUNDER

The performance of an echosounder designed for fish detection is most usefully expressed in terms of the greatest depth at which a single fish can be detected, though no manufacturer in his senses would put a bald figure on this without qualification. As with all other kinds of acoustic instruments, there are a good many factors affecting the signal-to-noise ratio, some of them depending upon the design parameters and others upon the operational and environmental conditions. Among the former are frequency, transducer size and efficiency, power, bandwidth and the thickness and material of the transducer window, if fitted. Among the latter may be included the size and type of fish being detected, the weather conditions, the size and speed of the ship and the position of the transducer. It is not enough to detect the fish, the signals must also be displayed in a form in which they can be recognised for what they are.

The fish traces seen on the echosounder chart come in a great variety of shapes and sizes: crescents, steeples, comets, plumes and thumbnails, are just a few of the many names that have been used to describe them. When a shoal that is too densely packed for the individual fish to be separately distinguishable passes under a ship, successive echoes take on a more solid form sometimes described as banks, walls and false bottoms. Too much attention perhaps has been paid to these shapes, since they are determined by the character of the echosounder and the movement of the ship quite as much as by the habits and behaviour of the fish, and virtually not at all by the actual shapes and sizes of the fish themselves. With a very directional beam each fish will only be detected once as the ship passes overhead. When the beam is less directional, several echoes are received both before and after the vertical position is reached. The curved trace is caused by the fact that the slant range is greater than the vertical distance (Fig 9).

Echosounder records ordinarily show a much compressed picture, so that banks or ridges on the bottom are made to appear much steeper than they really are, and the flat curve of approach of the typical fish trace is considerably foreshortened. This is how crescents become steeples and thumbnails become

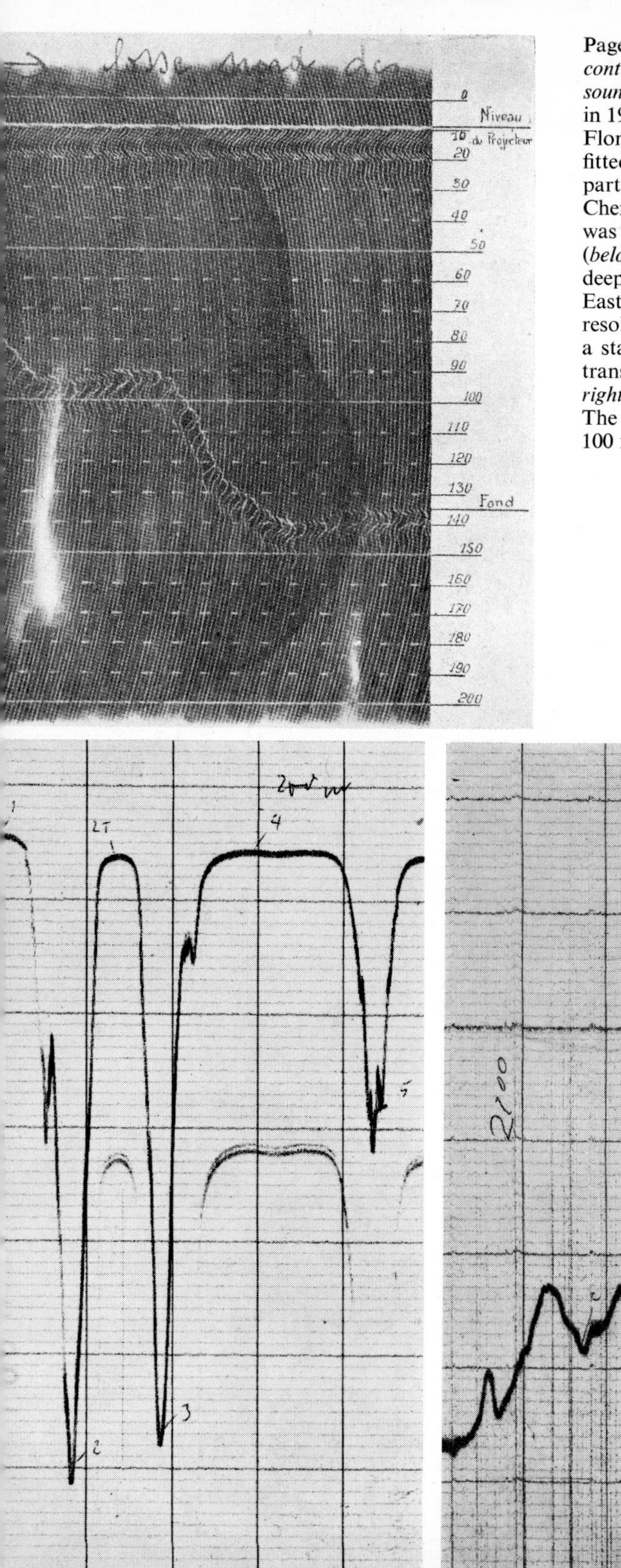

Page 85 *Bottom profiles with continuous recording echo-sounders.* (*left*) A record made in 1922 with the Langevin-Florisson apparatus (p 59) fitted in SS *Ile de France*. A part of the Hurd Deep off Cherbourg is shown. The ship was travelling at 23 knots. (*below left*) A recording of the deep-water canyon off the East African coast. The good resolution is the result of using a stabilised narrow-beam transducer (p 146). (*below right*) A sea mount or volcano. The horizontal lines are at 100 fathom intervals

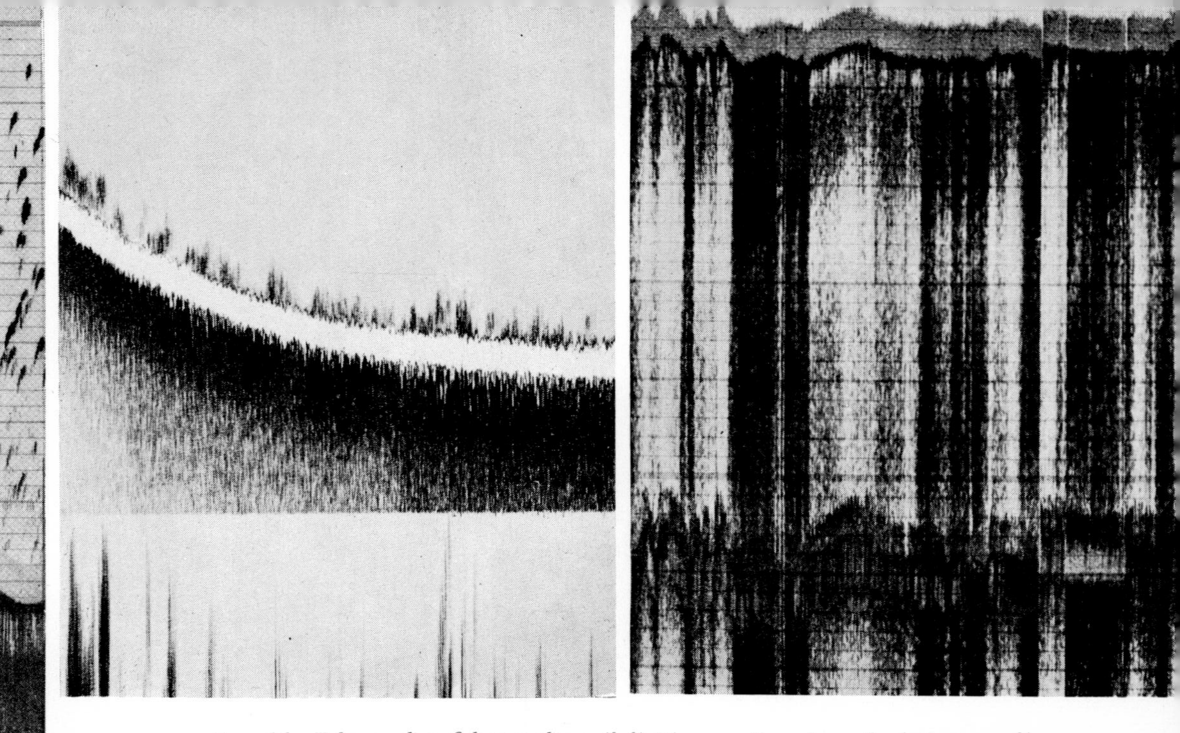

Page 86 *Echosounding fish recordings*. (*left*) The grey line above the bottom profile serves to show up fish echoes just above the bottom. The darker fish traces are of single cod. Depth 170 fathoms. (*above left*) Markings above the black edge of the white line are fish. The lower half of the chart shows an expanded, or magnified, picture in which the bottom has been 'flattened'. (*above right*) *Ground discrimination echosounding*. The black vertical strips are caused by rough or stony ground, which gives enhanced side echoes (p 95)

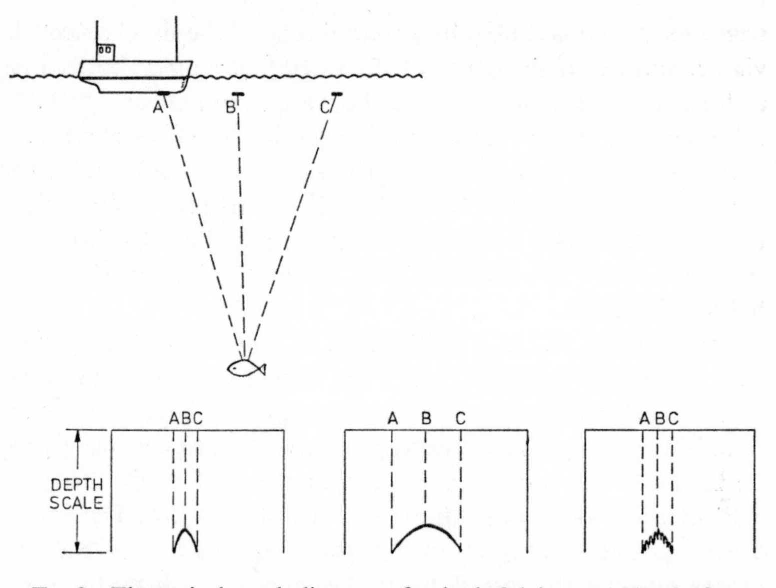

FIG 9 The typical parabolic trace of a single fish is caused by the fact that the sound beam is wide enough to detect the target before and after the ship passes overhead. The slant ranges to the fish from positions A and C are slightly greater than the vertical. The effect is accentuated because the echosounder chart compresses the record in the direction of the ship's travel. Of the three examples shown, the left-hand one is typical, the centre one is the sort of shape that is seen when drifting or trawling, and the picture on the right has a rough edge such as occurs in rough weather when the heave of the ship is added to the depth. All are exaggerated in size

comets. The speed of the ship affects the traces in a similar way and the heave factor gives them a rough edge. The characteristic shape is lost when there are several, or many, fish close together, as in a mid-water shoal. The effect is then of a solid paint, the upper edge of which accurately defines the depth of the shoal, while the lower edge can mark the chart far beyond the lower limits of the shoal, sometimes extending well into the bottom echo. This effect is not seen when shoals are detected by sonar, and a Russian fisheries research scientist, V. G. Azhazha, has suggested that it is caused by a second echo of the shoal reflected via the surface. If the depth of the shoal from top to bottom is at least as great as its depth in the water, the first and second echoes would merge into each other to give the effect observed.

It will be seen that all these various shapes are not related to the reflectivity of the fish themselves. Fishermen recognise in the traces on the chart the species they are searching for because they can relate the patterns they see to the known shoaling habits of the fish.

Special problems exist in the case of bottom-feeding species such as cod and haddock, which are caught on or near the sea-bed in depths up to 500m. Except when spawning they are never found in tight shoals. An average density of one cod per sq m on the bottom is exceptionally high and enough to fill the 'cod end' of the trawl within the space of a 40min tow. Detecting these fish individually in such depths requires a high performance set, and the task is made the more difficult by the presence of the much stronger echo from the bottom itself, lying immediately below the fish. The shape and duration of the transmitted pulse of sound is significant: a square pulse is ideal, meaning that the amplitude should rise immediately to maximum at the start of the pulse, stay there for the duration, and then fall equally rapidly to zero. This cannot be achieved in practice, but if the rise time is not extremely short, there is a tendency for the fish echo which is in the form of the pulse to be lost in the bottom echo, as can be seen from Fig 10.

Even when echoes from fish as near the bottom as this are detected, they are not necessarily identifiable as such on the recorder. Because of the relatively limited dynamic range of the recorder paper the fish echo, although in fact much weaker than the bottom echo, will nevertheless often be strong enough to saturate the paper; and as there is nothing blacker than black, the two echoes will not be separately distinguishable. This difficulty has been overcome by what is known as the 'white line' technique, which involves a simple amplitude gate in the

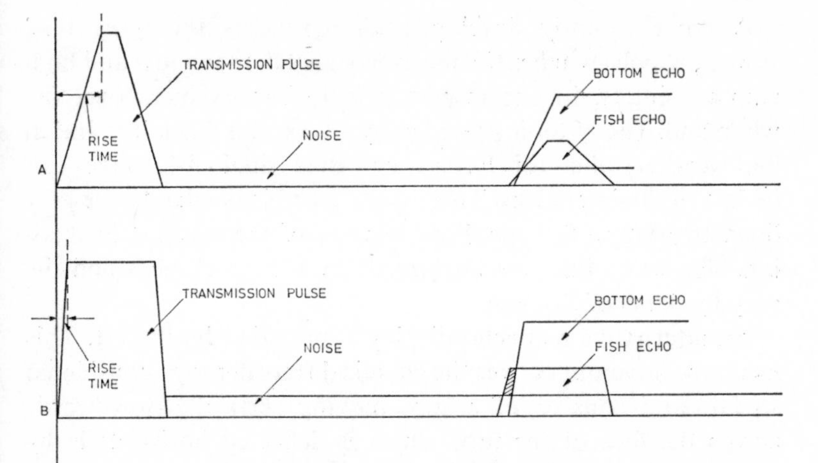

FIG 10 The two graphs of intensity against time are purely diagrammatic and not to scale. The 'rise time', ie, the time taken to reach the peak power from the start of the transmission, is repeated in all echoes regardless of their amplitudes. In graph A, because of the long rise time, the fish echo does not come above the noise level before the onset of the bottom echo and is therefore not recorded. In graph B the shorter rise time brings the fish echo above the noise level in time to be detected, as shown by the hatched area

receiver circuit to paralyse the receiver whenever the strength of the signal exceeds a certain value, the trigger point being set somewhere between the strongest fish echo and the weakest bottom echo ever likely to be encountered. The bottom echo, therefore, but not the fish echo, cuts out the amplifier, so the first part of the former does not mark the chart and leaves a gap in the trace; successive gaps build up into a white line that follows the contours of the bottom and serves to separate the fish echoes from the seabed echoes. The idea was born when a service engineer noticed that the amplifier in a fishing echo-sounder was overloading due to maladjustment. The possibilities of this chance mutation were realised and in due course turned into a beneficial modification which has, over the years, become a standard fitting.

A more recently developed alternative is the 'grey line' system, which switches the received signal between low and high gain amplifiers. In one respect it is an improvement over the white line. The latter is preceded by a very thin black line and in bad weather, when this has a rough edge, single fish echoes can be lost in the interstices. There is no such black edge to the grey line and bottom fish stand out because of the contrast in shading. The white line was introduced by Kelvin Hughes and the grey by Atlas of Bremen.

An alternative fish echo display is provided by the crt. This has some advantages over the chemical recorder type considered so far, and consists of a spot moving vertically downwards across the face of the tube which is deflected horizontally by echoes as they are received. There is no limit to the speed of travel of the spot and, therefore, to the extent that any particular section of the water column between the surface and the seabed can be expanded or magnified. It is commonly known as the loop, derived from 'Fischlupe', the German for fish lens, whence the idea originated. This type of display has a considerably greater dynamic range. The bottom echo has many times the amplitude required to reach the edge of the tube, and its writing speed is in fact so high that it is scarcely visible at all. This again helps to distinguish the echo from fish just above the bottom, which makes a brighter line because the spot is moving more slowly. But this kind of display has no memory, and so the skipper must watch it constantly. Modern fishing echosounders normally provide the option of having both types of display. One with a crt display only has the advantage of a continuously adjustable prf.

Although most of the sound energy transmitted into the water is concentrated in one direction, the spreading of the beam to the half-power points is such that at a depth of say 100m or more the area of the seabed that is insonified by the sound pulse is quite considerable. Fig 11 shows that any fish lying within the shaded areas will not be recorded, because the echoes from them

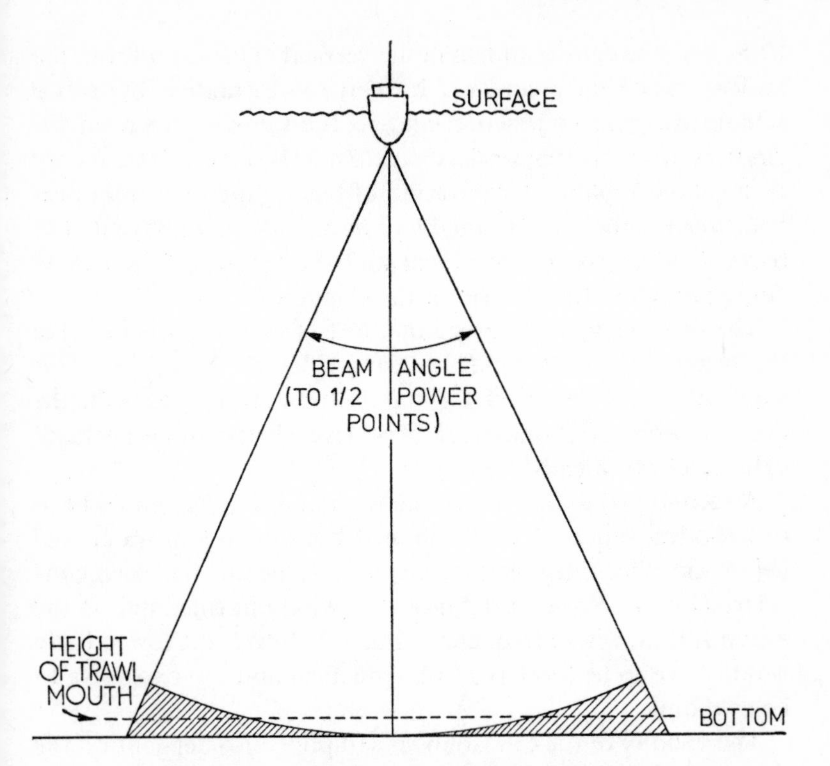

FIG 11 The width of the beam can result in misleading information concerning the density of fish on the bottom. Fish lying within the hatched zone will not be detected, as the first bottom echo will already have been received. On the other hand, fish lying just above the zone near the limits of the beam will appear on the chart as though they were closer to the bottom than they really are. The dotted line represents the height of the trawl headline. If, in fact, the fish are all above this line, and evenly spread, it will appear on the chart as though they extended right down to the bottom and the catch would not measure up to the expectation

will not reach the ship until after the much stronger bottom echo from the closest point. Moreover, the fish near the edge of the beam but lying just above the shaded areas will appear to be close to the bottom, whereas in fact they are some way above it.

If the average density of fish in the vertical plane is uniform, the indications on the record will be truly representative, but this is seldom the case in practice, and so the sample shown on the chart is to a greater or lesser extent misleading. The record could show a good concentration of fish right down to the sea-bed, when in fact the fish might all be at least 10m above it. The trawl would pass beneath them and the net would be hauled empty, much to the chagrin of the skipper.

The obvious way to reduce this sort of ambiguity is to make the beam more directional by using a larger transducer. This step introduces the need for stabilisation, which, as with the oceanographic echosounder, is expensive whether done mechanically or electronically.

A second possibility, at once cheaper and simpler, would be to tow the transducer in a streamlined housing in still water well below the ship. This idea, attractive in principle, has been considered by the White Fish Authority (WFA) but ruled out on the grounds that, with a frequency of about 30kHz, the towed body would have to be too large to be practical and too expensive to be economic.

The validity of the echosounder sampling also depends on the assumption that the trawl, perhaps as much as a mile behind the ship, is in fact directly astern. However, in the presence of a strong cross wind the ship will be blown down to leeward and the net will be on the quarter.

The logical conclusion to all these difficulties is to abandon the echosounder altogether, and to start again with a transducer mounted on the headline of the trawl and pass the information to the ship either by cable or acoustic link. The problems of acoustic telemeters are considered in relation to their various applications in Chapter 6.

Since World War II echosounders have been used in ever-growing numbers to confirm the presence (or absence) of fish and to give a purely qualitative judgement on the size and density of shoals detected. More recently there has begun to be a need for quantitative measurements. The fisheries research worker needs to put figures to the results of his abundance surveys, and the trawler captain wants to know the catching rate as precisely as possible in order to decide when to haul. Experimental equipment has been devised to count the fish, both in mid-water and on the bottom, but from what has gone before it will be obvious that the former is an altogether simpler task than the latter.

The simplest way to make a quantitative assessment of fish stocks applicable to mid-water fish with well defined shoaling habits involves no more than a count of shoals detected. By making an assumption for average density and multiplying this by the measured dimensions of the shoal, a first approximation of quantities can be made. No special equipment is needed for this, but the method is obviously crude. As already explained, the lower edge of the shoal cannot be reliably estimated from the echosounder chart, which in any case only displays a slice through the middle and gives no indication of width.

A much more precise method is to count individual fish echoes and, by recording echo size distribution, to estimate population density and the distribution of fish sizes. It will be seen from the earlier remarks about the relation between fish traces and echosounder characteristics that this method, potentially capable of considerable accuracy, does involve a precise knowledge of the directivity function of the transducer. To get the best results, multiple echoes from the same fish must be eliminated as far as possible, which involves the use of a very directional sound pulse derived from a stabilised transducer of small dimensions and limits the technique to a high frequency and shallow water. This method of fish counting has been developed and successfully used by the Marine Laboratory at Aberdeen.

For marine biologists and other fisheries research workers even more perhaps than for fishermen themselves it would be an advantage to be able to take the echosounding art a stage further and proceed from fish counting to fish identification or, at least, estimation of size. Much painstaking research aimed in this direction has been carried out in recent years. Dead fish have been impaled on the point of a darning needle and acoustic back-scattering patterns have been traced in the form of polar diagrams about the vertical and horizontal axes of dead fish with onazote for swim bladders.[2] These have been compared with various geometrical shapes such as ellipsoids, air-filled cylindrical holes and so on, that are more amenable to calculation than the works of nature. Variations of reflectivity with frequency have been investigated and a great many model experiments have been made, using scaled-up frequencies in the MHz range and scaled down fishes such as sticklebacks.

The results of these endeavours are inaccessible to the layman and as yet of no significance to the fisherman. About all that has been achieved is a confirmation of the importance for echo detection of the gas-filled cavity, possessed by most species, known as the swim bladder. Though representing only about 5 per cent of the total volume of a fish, this swim bladder is responsible for 50 per cent of the returning echo. Species such as mackerel which do not have a swim bladder are less reflective than most others, such as cod, hake, haddock and herring, which do. Identification of fish species by echo is indeed an intractable problem and it seems unlikely that a way of doing this by using active, or echo-detection methods, will ever be discovered. Passive detection, ie, listening to the sounds that the fish themselves make, might lead to positive results one day. Unfortunately the species in commercial demand are among the more silent majority.

The state of knowledge of this subject, marine bio-acoustics, is discussed at the end of Chapter 6 (p 185). Some workers in this field have concluded that for certain species the swim bladder acts as a vlf sound resonator; it is thus an important mechanism for active and passive detection alike.

GROUND DISCRIMINATION

Echosounders are used by fishermen not only for detecting fish. Danish seine netters rely on them for information about the nature of the bottom. A Danish seine (not to be confused with a purse seine, which is a surface net for catching mid-water shoaling fish such as herring) is a kind of lightweight bottom trawl. Instead of being towed along the bottom, it is laid out in a suitable area and then hauled up to the ship. It is quite easily snagged and damaged by, or lost on, wrecks, large rocks and other obstructions on the sea floor. On the other hand flat fish such as plaice and sole are often found near the edge of areas of stony or rough ground. Fish of these species lying within 2 or 3in of the bottom are virtually impossible to detect with today's instruments, though largely unsubstantiated claims to have done so are made from time to time. So it is the bottom echo that the fisherman is watching for when standing by to lay out the net. A smooth bed consisting of hard mud or fine sand returns a short echo of much the same length as the transmission pulse itself, whereas rougher ground made up of stones and gravel gives a much more extended echo. As soon as these tails appear on the trace, the skipper puts the wheel over and slips the net, which will then lie along the edge of the rough ground. The echo tails are formed in much the same way as the wings of the typical crescent echo trace of a single fish. There is no significant difference between the reflectivity of smooth and rough ground but the back-scattering coefficient for the latter is proportionately greater. This means that when the beam strikes the bottom obliquely, a stronger signal comes back to the ship from the rougher ground, an effect accentuated when the main lobe and the secondaries of the transmitted pulse cover a relatively wide arc.

The way in which special transducers came to be developed to accentuate the difference between rough and smooth ground is another example of 'feedback' development. From the explanation already given it will be evident that it is the energy transmitted outside the main beam that is significant in producing the longer tails to the bottom echoes. In earlier models using ring transducers and conical reflectors the beam patterns were, in fact, none too directional. However, the flat-faced transducer concentrates more of the energy into the main beam, a desirable feature for most requirements, but not for this one. Fishermen who installed the new equipment were not slow to complain that its ground discrimination qualities were actually worse than before. The reason was quickly identified and a measure of 'anti-tapering' was introduced to bring ground discrimination standards back to what they had been. This was done by filling the centre section of the transducer with inert stampings.

Some research into the history of Marconi's developments shows that ground discrimination is not a new thing. In the 1930s they produced an echometer known as Type 414, which had a piezoelectric transducer of the Langevin-Florisson type operating on 37·5kHz. The depth was indicated by a spot of light, reflected from the mirror of an oscilloscope, which travelled horizontally across a semi-transparent curved scale about 1ft long. The spot of light was deflected vertically on transmission and by echoes received subsequently. It was thus a kind of optical analogy of the fischlupe, and it was said that with its aid, after a little practice, even the nature of the seabed could be distinguished.

'Spoiling' the transducer beam angle is not the only way in which the contrast between bottom echoes in smooth and rough ground can be accentuated. The use of a ceramic transducer in place of the magnetostriction type is helpful simply because it is more efficient. An example of ground discrimination achieved with a standard barium titanate transducer is shown in the plate on p 86. In this chart, which is typical of many obtained with modern echosounders, it will be evident that the explanation given above to account for the longer tails in rough ground is inadequate by itself to account for the very marked differences.

It may be that some of the sound follows multiple reflected paths between the larger particles making the rough ground before finally being back-scattered to the ship.

HORIZONTAL DETECTION

Throughout World War II any number of promising echoes obtained by anti-submarine escort vessels while searching for U-boats turned out to be fish shoals, dolphins, whales or other marine mammals, to the disappointment of the operators and the discomfiture of these peaceful inhabitants of the sea. The tense excitement and subsequent let-down is well expressed in the following piece by J. Renou:

> 1943—Un convoi venant d'Amérique traversait l'Atlantique a déstination de Gibraltar. De chaque côté, devant, derrière, les escorteurs patrouillaient. Sur les pasarelles on entendait le 'ding', 'ding' monotone des émission de l'asdic, à la recherche du sous-marin ennemi. Aucun écho. Alerte. Echo au 300. Distance 1,000 yards, signale le Sabre qui part en chasse.

and so on through several paragraphs of breathless suspense until:

> Le convoi anxieux attend qu'il exprime son opinion. S'agit-il d'une meute qui guette?
> Et, simplement, une voix très calme explique:
> —Le but detecte était un banc de poisson.[3]

The author goes on to describe a series of trials set up by the French Navy in July 1946 to investigate the possibilities of putting such sonar echoes, unwanted in wartime, to good effect in peace. After some disappointing results success was achieved in the autumn in the Bay of Biscay, when good echoes were obtained from sprat shoals near the surface at a range of about 1,000m.

That was 25 years ago and one might have supposed that the significance of these results for commercial fishing would have led to a rapid development of sonar especially designed for fish detection. Such, however, was not the conclusion of a report on echosounding and asdic for fishing purposes called for by the International Council for the Exploration of the Sea (ICES) some years later. The report, published in 1955, was based on the answers to a questionnaire circulated to all member countries. The preface stated that the replies showed sonar was only being used for whale tracking and as yet had no place in other commercial fisheries except the Norwegian herring fishing industry, where it was being tried out.

For the sake of chronology, therefore, it will be convenient at this point to break off the narrative of the development of sonar for fish detection for an excursion to the Antarctic in search of whales.

WHALE TRACKING BY SONAR

A whale, when it is beam on, returns a strong echo at ranges of 1,000m and more. No doubt there were many people whose wartime experience of these 'non-sub' echoes gave them the idea that sonar might play a useful part in the business of whale catching. But it was Jock Anderson, Chief Scientist at the Admiralty Underwater Detection Establishment (UDE), who authorised the first attempt to apply sonar to the whaling problem.

A trial was undertaken during the 1945–6 Antarctic whaling season, with two whalecatchers that had been fitted with sonar for war service. The scientist on the staff of AUWE selected for this task was W. J. McCarthy, and his report was made available to Kelvin Hughes, which in due course turned it to good account commercially. The objects of the trial were defined as follows:

> To determine the usefulness of the asdic equipment provided in the detection of and maintaining contact with submerged whales.

If successful in this object to

> (i) determine the procedure which should be used when whaling.
> (ii) determine the type of equipment which should be provided for future asdic-fitted whalecatchers.

The results, summarised in essence in the following paragraph, also taken from the report, were essentially negative:

> Under favourable conditions asdic contact with whales was obtained and held at ranges up to one mile in telephones and 1,200 yards on the recorder. The harpoon gunner's chief requirement, however, is for holding contact at short ranges (down to 10 yards) and in this the asdic equipment fitted failed to give material assistance.

Whales are first sighted at a range of anything from 5 to 10 miles by the 'blow' made visible by the vaporisation of the relatively warm air expelled from the lungs as the animal surfaces to take another breath. This is far beyond the range of the sonar, which therefore plays no part in initial detection.

Having sighted a school of whales at a distance, it is the gunner's object to approach them silently so as to take at least one unawares. Thereafter the stalk becomes a chase. Fin whales can attain a speed of 14 knots, about equal to the maximum available to the prewar catchers, but usually tire within half an hour or so. Blue whales on the other hand seem able to swim at such speeds or even greater for an indefinite period. In a catcher capable of 17 knots I once watched a blue whale as it maintained its stately progress, keeping a steady 200yd ahead of the ship, for several hours, until the frustrated gunner eventually abandoned the chase. Rough weather, of course, will slow the ship more than the whale, as well as making an accurate shot more difficult. On the other hand, when swimming fast these whales lift their heads higher in the water when they come to blow, presenting a slightly larger target and giving the gunner perhaps an extra second to take aim and fire. If the whale is swimming on a steady course, the gunner has no difficulty in keeping it on the bow until he can come within range (10–60m according to preference) for a shot. In such circumstances the final outcome is scarcely ever in doubt. Even a miss (causing groans from the crew) and time lost in reducing speed to haul in the forerunner and harpoon is not the end of it. The whale, foolish creature, believing the danger to be over will at once slow up to allow the chase to be resumed with little extra distance to be recovered. Once in a while darkness may intervene to terminate the hunt, but since, in the Antarctic latitudes where the hunting occurs, it lasts only from 10 pm to 2 am, this cloak is inadequate enough. Sometimes the whale, especially the fin, will alter course between blows in an irregular way by as much as 45°. Such whales, known as 'crazy' (though exactly the opposite from their own point of view, did they but know it), set the gunner an almost impossible problem, the more so for the fact that such changes of direction occur when the whale has completely submerged after a blow. The gunner can steer towards the last known position only to find the whale next time appearing on the beam as far away as ever; or, using his hunch, he is as likely as not to find himself completely wrong-footed. It was in this predicament that sonar was thought of as an aid to whale tracking.

The technical difficulty of improving on the explosive harpoon invented by the Norwegian explorer Sven Foyn in 1868 is formidable. Electric harpooning has been tried unsuccessfully. It would be possible to combine the sonar with a gyro-stabilised sight and a modern weapon system, but since the present method does work, the heavy development expenditure that would be involved could only be justified on humanitarian grounds. In any case, a more humane weapon could only be introduced with the co-operation of the gunners themselves, and this has not been forthcoming, since they have a vested interest in the great skill needed to use the old Sven Foyn gun.

The gunner of *Southern Wilcox*, the catcher with the more promising type of sonar and the one in which McCarthy sailed was unsympathetic from the start. The ultrasonic ping would, he believed, disturb the whale and so do more harm than good. In the event it became evident that sometimes, though not invariably, the sonar transmissions did have the effect of making the whales in the vicinity increase speed. Another objection, of greater substance, was the risk of fouling the forerunner round the streamlined dome that houses the transducer. In this design there was a narrower neck above the dome itself which in the lowered position protruded beneath the hull. When the harpoon has struck home the forerunner, coiled up beneath the gun platform, continues to run out so that bights of slack rope form in the water and the ship's engines are immediately stopped to avoid the risk of fouling the propeller. Being mounted near the bow the asdic dome represents a correspondingly greater hazard. On one occasion during this trial the forerunner did foul the neck of the dome but the ship was put about in time to clear it. Failure to do so could well have damaged the dome or parted the line as the strain came on when the whale attempted to escape by sounding.

Quoting once more from McCarthy's report:

> It was therefore imperative to house the dome before the harpoon gun is fired, thereby imposing a severe limitation on the use of the Asdic gear by interrupting contact at the critical point of the chase. In practice, the dome must be housed whenever whales surface within harpooning range, in anticipation of the gunner deciding to fire. Should he

> decide not to fire, the order to lower the dome is passed after giving the whale time to dive. The time-lag in lowering the dome, renewing Asdic search and gaining contact may amount to 1½ minutes.

Faced with so fundamental a limitation the reader may wonder how it was possible to obtain the many excellent asdic traces of whales contained in the report, of which an example will be found in the plate on p 103. The answer is simple.

The position of the dome was indicated by lamps on the bridge control unit—green when it was raised and red when it was lowered. Mr McCarthy, not the sort of man to go all the way to the Antarctic to be frustrated, took the precaution of wiring in another switch, hidden from view, with which he could override the lamps. So, if caught operating when the gunner had told him not to, he could quickly switch to 'green'. He also took the precaution of omitting to mention the fact in his report.

A full understanding of the purely technical problems would involve a description of the types of wartime sonar fitted in the two vessels, but these sets are long since obsolete and their particular limitations are of little interest today. Suffice it to say that the main difficulty was associated with the methods of training the transducer. In *Southern Wilcox* both methods of training were fitted to enable their efficiency to be compared. One allowed for training relative to the ship's head, and the other independently of the ship's head by means of a line of light on the compass card in the binnacle. The heavy yawing in a seaway, which might amount to as much as 20°, together with the continual changes of course ordered by the gunner during the chase, made the relative training method virtually useless. On the other hand, when training by compass bearing the operator is independent of the bridge and cannot visualise the relative picture. This way was the better but still far from ideal.

The value of sonar only becomes significant at 500yd range and increases as the range is reduced. So we may now consider

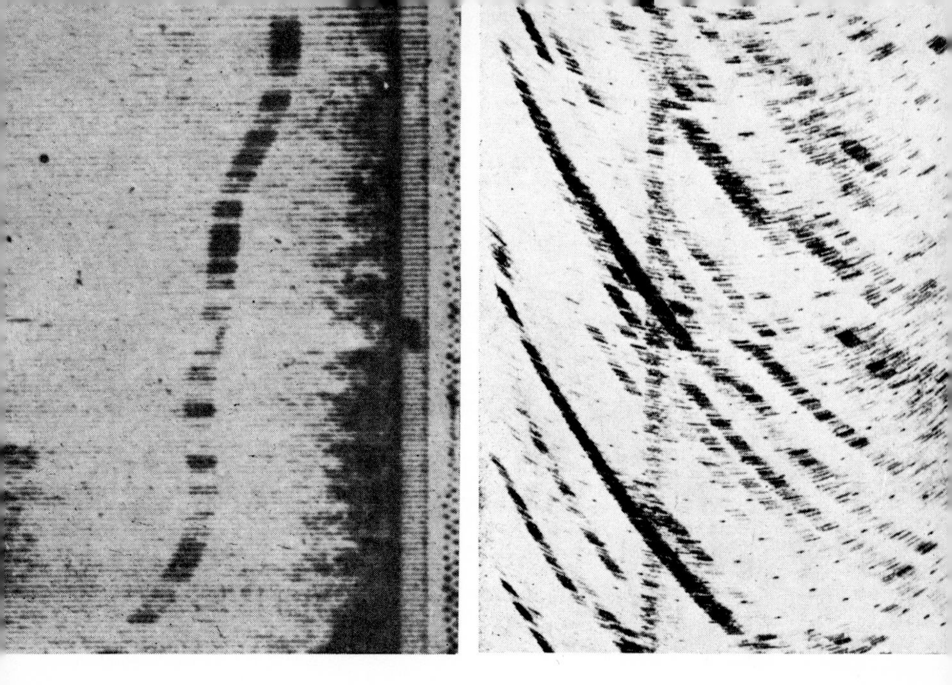

Page 103 *Sonar records of whales, porpoises and ice*. (*above left*) Strong echoes from a fin whale on the beam. (*below left*) Much weaker whale echoes are received from the astern view during a chase. The blacker markings are blow wake echoes which, being stopped in the water, are closed much more rapidly (p 105). (*above right*) Short range echoes from porpoises off the Norwegian coast. (*below right*) Icebergs make excellent sonar targets. This one was first detected at 3,500yd (3·2km). Such an instrument might have saved the *Titanic*

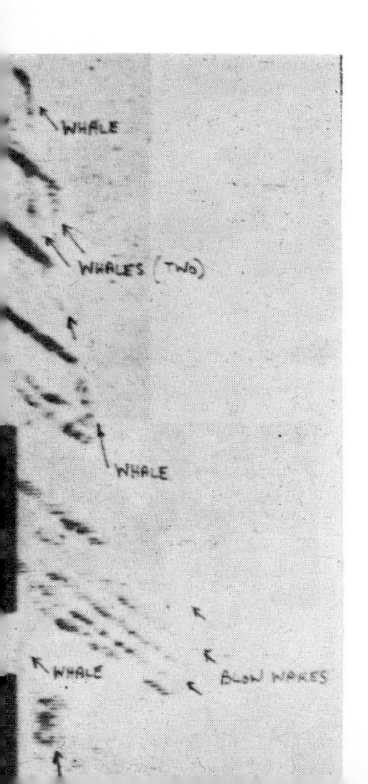

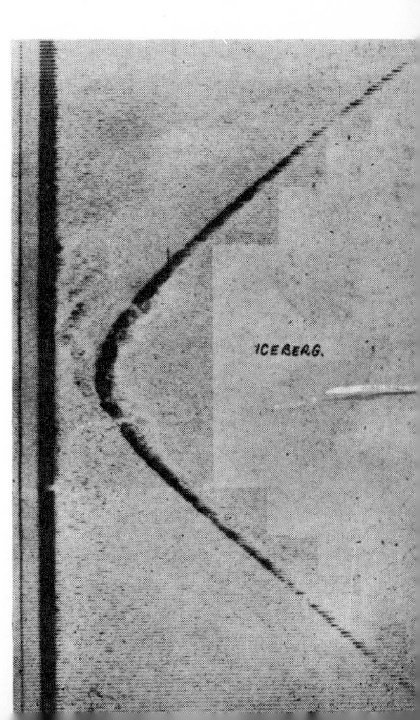

Page 104 *Dual-frequency echosounding*. (*above*) The dark trace (30kHz) shows the bottom of a silt-filled trench while the lighter trace (210kHz) shows the top of the silt layer (p 128). (*below*) In this example the connections have been reversed so that the dark trace now indicates the low-frequency echoes and the light trace the high-frequency ones. This has made it possible to distinguish genuine peaks from sidelobes in an area of steep coral outcropping. Because the high-frequency transducer has a narrower beam, the dark trace shows the true profile, whereas the light grey protrusions, coming from the wide beam low-frequency transmissions, are sidelobe echoes (p 131)

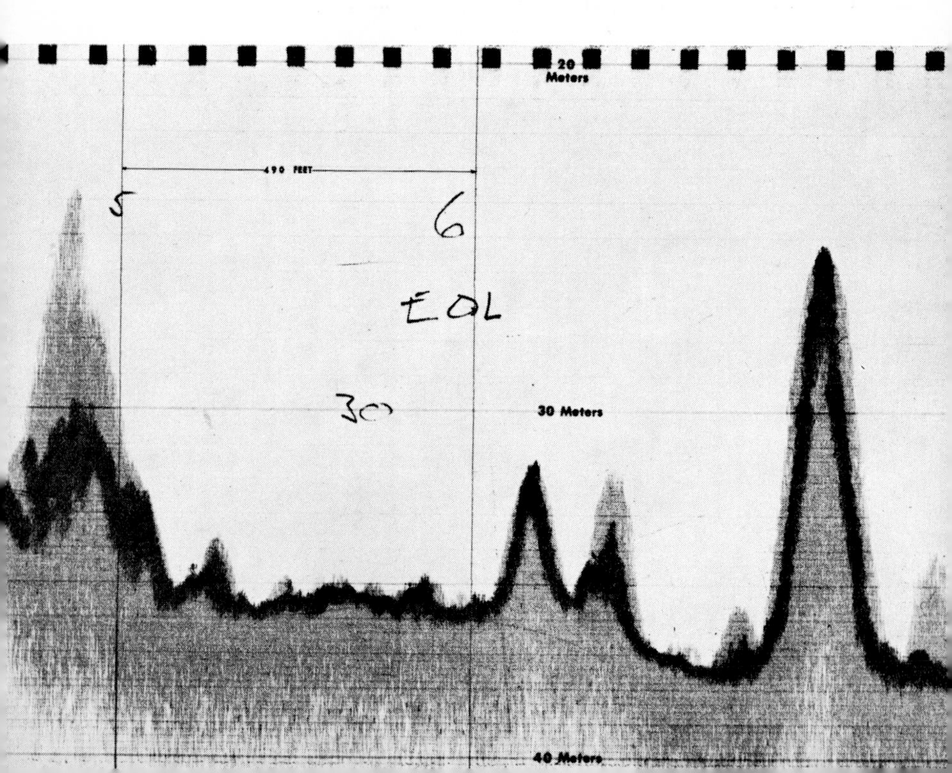

the operational problems posed by the whale itself as an echo target and by the difficulties of working at very short ranges. These sets were designed for submarine detection when a U-boat was limited to a speed of 2–3 knots underwater; the operator, having obtained a promising contact, could take his time in deciding whether or not the echoes were indeed from the sought-after target, while lying off at a convenient range before moving in to the attack. In whaling, however, the approach develops into a stern chase, from which direction the whale presents a very weak target because of its streamlined form. Nor does its movement through the water produce any wake or hydrophone effect such as is typical of any propeller-driven vehicle. The one characteristic of the echo that helps the operator, without which the detection of a whale from behind would be virtually impossible, is the 'doppler' effect. This frequency shift is so marked that an experienced operator can recognise the whale echo even when its amplitude is actually less than that of the background signal. This explains the advantage of headphones over the recorder. However, by way of compensation, the whale enjoys a measure of protection from the circular patch of aerated water it leaves behind after each blow. This blow wake, as McCarthy christened it, is reminiscent of the submarine bubble target (SBT) already referred to in Chapter 2, though it is visible on the surface and, of course, has no doppler. None the less, when it is between the whale and the ship, it has the effect of masking the former completely; the whale is always lost momentarily after it has surfaced to blow. It will be evident that a school of whales between them leave behind a more or less impenetrable acoustic barrier, and when the ship is separated from its prey by more than the distance between successive blows, sonar detection is virtually impossible.

At 600m the blow wake subtends 20° or more, compared with 3° for the whale itself. Moreover, when the ship steams through a blow wake, the aeration is sufficient to quench the signal completely. The cause of the blow wakes is rather obscure: while it would be understandable for the whale to start exhaling before breaking the surface, this can hardly be the case when it is on the way down; and yet the blow wake is certainly not produced until the end of the surfacing. Records indicate that whales sometimes make blow wakes when submerged. McCarthy suggests that an explanation of this may be related to their method of feeding, although one would hardly expect a whale to take time off for a mouthful of food when being chased. Whales of the *Balaenoptera* species feed exclusively on a type of shrimp a few centimetres in length called *Euphausia superba*, commonly known as krill, which are found in their millions in surface shoals in the Antarctic. These large mammals require a correspondingly substantial diet to keep going. When moving into a dense shoal of krill, they take in a great mouthful of sea and, by contracting the throat muscles and with the aid of the tongue, then expel the water through the spaces between the whalebones, the inner fibres of which act as a filter in which the krill are trapped. It has been estimated that a blue whale can catch as much as a ton of food in a single swallow. Ironically, man, having almost succeeded in exterminating the species, is busily devising man-made traps to catch the krill. He is unlikely to match nature's own elegant solution.

To return to the blow wake, this is more likely to be caused by the fact, already mentioned, that when the whale is being chased, much more of the head is raised out of the water when blowing, the mouth appearing to be slightly open. On diving, this intake of air would be expelled and so form the patch of aerated water that has proved so inconvenient to the pursuer.

Ideally the gunner is looking for information on the whale's position underwater down to the point when it comes within firing range. At these very short distances the echo follows so quickly after the transmission that eventually it is lost in the blank-off period of the receiver. Shortening the pulse length of the transmission has the undesirable effect of making both the echo itself and the reverberations against which it is heard more noisy, and less musical, so that the frequency shift becomes more difficult to detect. About the shortest pulse that has any musical sound in it at all is 10ms, and even this represents an echo distance of 7·5m in the water. It is therefore just possible for a skilful operator to hold contact down to firing range.

Fin whales typically appear above the surface for 2–3sec at intervals of between 5 and 30sec. When undisturbed, this cycle is interrupted periodically by a much longer period below the water, which may last for as long as 10min. This is known as sounding, and the whale may then dive to as much as 100 fathoms, though this is purely conjectural and is based on the ranges at which contact in the sonar beam is lost. When chased, these whales sound less frequently and sometimes not at all. Blue whales show similar behaviour patterns but tend on the average to blow less frequently. When they are chased, intervals between 15 and 60sec are typical; but periods of diving, surfacing and sounding recorded by McCarthy show wide variations from the norm.

The habits of the toothed sperm whale, of which only the male of the species is ever seen in the Antarctic, are markedly different and they are relatively easy to take. Even so, sonar was used to track them, despite the fact that these whales, when harpooned, had a greater tendency to turn back and foul the line round the dome.

The conclusion then that must be drawn from this first attempt to apply the technology of underwater sound to that of catching whales is that it bristles with difficulties and is in many respects more exacting than the comparable operation of finding and attacking submarines. It may seem surprising, therefore, that Kelvin Hughes, which was probably at that time the leading commercial manufacturer of echosounders in the world and possessed the know-how to adapt its techniques to horizontal detection, should have chosen first to tackle the very difficult whale-tracking problem in preference to the much easier task of detecting mid-water shoals of fish. The explanation lies in the fact that it was then generally believed that the price of sonar would prove to be beyond the means of the fishermen who make their living from mid-water shoaling fish such as herring, pilchards and sprats. Big business in British fishing has always been with the near and distant water trawlers that search exclusively for bottom-feeding fish such as cod and haddock. No way has yet been found to detect these species with a horizontally or diagonally directed beam.

In 1949 negotiations for the development of a sonar suitable for whale tracking were opened between Kelvin Hughes and Christian Salvesen & Co of Leith, which was then operating two whale factory ships with their attendant fleets of catchers, in addition to a shore station and base at Leith Harbour in South Georgia. Against all the odds they were successful. The Type 128 asdic, of which the Admiralty still held substantial stocks surplus to requirements, was used as the basic design. The amount of development involved was in fact relatively limited, the major units of the service design, such as the raise/lower gear, dome, control training unit and amplifier remaining virtually unaltered. The purchase of these surplus stocks ahead of the scrap merchants, however, was not achieved without difficulty.

A new transducer of the magnetostrictive type, having two frequencies (25kHz and 14kHz) back to back, replaced the Admiralty quartz design. Two of the main recommendations of the McCarthy report were adopted: the neck of the dome was filled in by a wooden fairing to reduce the risk of fouling the forerunner, and as a further precaution a hydraulic lifting gear was developed to raise the whole ponderous apparatus in a magical 2½sec instead of 1½min. This latter, however, was not introduced for several years, by which time sonar had already proved its value and was being widely fitted. The first Kelvin Hughes trial took place in the 1950–1 season, and the operator who accompanied the ship had one enormous advantage over McCarthy, in addition to the technical improvements in the equipment itself, in that Salvesen, having decided to go ahead with the idea, selected a gunner who was already a sonar enthusiast. He was Captain Ansgar Flaaten, who had commanded a trawler fitted with sonar during World War II. Should he ever happen to read these words, I hope he will forgive me for mentioning that he did not happen to be one of the top gunners. His place in the league table had been consistently near the bottom. The top gunners in the Salvesen fleet had no objection to this arrangement, since, at that time, they did not believe in the equipment anyway.

The whaling season was, and what little is left of it still is, controlled by international agreement. Catching by factory ship expeditions starts at midnight on a certain date early in the New Year, ie, in the middle of the Antarctic summer, and ends when all the expeditions have caught between them a previously agreed total. It is a race against time. By the appointed date all the factory ships were ready for the 'off' in whatever particular area they had selected, with their catchers ranged about them. At 5min past midnight Flaaten reported by radio to his factory ship *Southern Harvester* 'whale fast'. This was unprecedented. In the half light of the Antarctic night a 'blow' could not possibly be seen further than 100yd or so. The obvious conclusion that he had shot the whale the previous evening and had delayed his report was, however, not the correct one. The whale had indeed been sighted earlier and then followed in the dark for the 2–3 intervening hours by the sonar operator before being harpooned as soon as the midnight deadline was passed. The other gunners of the fleet took note. Perhaps because whales were relatively scarce, Gunner Flaaten, with the aid of his sonar, actually achieved top billing for the first 3 weeks of the season. Then the generator broke down and, as this was a prototype equipment, there was no spare onboard, nor any possibility of effecting repairs on the spot. This, however, turned out not to be an unmitigated disaster. True, it brought the experiment to a premature end, but when in the ensuing weeks this gunner, robbed of his special aid, began to drift back to end the season more or less in his usual place near the bottom of the league, where a gunner faces relegation to captaincy of a mere whale-collecting boat, the conclusion could no longer be in doubt.

Thus was the climate of opinion reversed. Top gunners were in the habit of getting what they wanted, and this was now sonar.

Inevitably there were further problems, and the gunners who were now clamouring for this new device soon found that it was not all as plain sailing as they had imagined. Details of operating ranges and times between blows by themselves fail to convey the sheer speed and skill with which the operator has to make up his mind about the validity of the echoes he could hear. The bridge control unit had by then been put in a perspex hut above and behind the bridge, and a newly installed communication system enabled the operator to broadcast his conclusions to the gunner, whether he was on the bridge or the gun platform. The operator had the advantage of an uninterrupted view and could always see the whale as soon as it broke surface, as well as the helmsman and the gunner. During the second or so after the whale disappears, perhaps 200yd or less ahead of the ship, the echo is totally masked by the blow wake. But already the gunner is gesturing to the hut to know which way to turn; he is not concerned with technicalities but just which way the whale has gone. The operator searches to either side of the powerful and rapidly closing blow-wake echo, listening for that faint low note. There's something there but can he be sure? Never was there a truer saying than, 'He who hesitates is lost.' The echo is still unconvincing but he has to make up his mind and take his chance. 'Whale going to starboard—100 yards.'

The gunner, by now on the platform, waves to the helmsman and the ship swings round. The transducer is locked to the compass card, except when the operator overrides the control and moves it a little to one side or the other. Even so, as the bow swings rapidly right or left in response to the wheel and maybe to the sea as well, the operator has his work cut out to hold contact. 'Steady now, port a little,' and by then, if he is right, the whale surfaces again just under the port bow and the gunner is ready with the harpoon on bearing. But if he is wrong and has been chasing a shadow, woe betide him, for then the whale will surface away off on the starboard beam and the language of the gunner will be unprintable.

So the operator had to be someone able to make up his mind quickly and yet not be stampeded into a wrong decision. To assist in the selection of the most suitable candidates a training unit which produced a good simulation of the real thing was designed, even to the extent of a bridge unit mounted on a rolling platform.

The gunners came to like the sonar as much as anything for the fact that it enabled them to limit the time spent on the very exposed gun platform in the bows of the ship, and to have the harpoon pointing in roughly the right direction before the whale surfaced.

Virtually all the whalecatchers of the Western whaling fleets, mainly British and Norwegian, were fitted with sonar in the end, as well as a number of those operating from shore stations in South Georgia and South Africa. The Japanese bought a few and then, predictably, made their own 'Chinese' copy. The Russians showed keen interest, but since it was considered that these old sets still had some wartime potential, an embargo was rigidly imposed. Only when it was possible to prove that they had themselves designed similar equipment, were the various government departments concerned persuaded that sales might be permitted. By then the market had gone.

HORIZONTAL DETECTION OF MID-WATER FISH SHOALS

The delay of 10 years or more between the publication of the results of horizontal echo-detection experiments made immediately after World War II and their commercial application was caused by economic factors. The detection of mid-water shoaling types of fish presents fewer and less severe operational problems than those associated with whaling, just described. On the other hand, the financial considerations are more significant. As explained, the problem of tracking whales has marked similarities to that of attacking submarines, so that the equipment used during the war was broadly adequate for the task. Fortunes had been made in whaling before World War II, and when the companies continued their operations after it was over, there was money available for such new equipment, costing £10,000 or more, in addition to the back-up services of operator training and maintenance. Figures of this sort, for an instrument that still had to prove itself in the commercial field, were then beyond the resources of the mid-water fishing industry. Furthermore, the main underwater components of the type of sonar used during the war were far too bulky for installation in the much smaller vessels used for this type of fishing. Consequently a completely new kind of design was needed, and this would have to be developed without the aid of government money. Another inhibiting factor, particularly applicable to Britain, was the essentially passive methods in general use for catching herring. The technique mainly used was the drift net, laid in long lines and buoyed at the surface to form a vertical barrier into which the fish would swim and be caught by the gills. Great expertise, handed down from one generation to the next, was needed to choose the place and the time to lay the nets, and success in this method depended more on a knowledge of the likely behaviour of the fish and the conditions in which they could be expected to rise to the surface and shoal in dense concentrations than on the use of echo detection.

New methods of catching mid-water shoals were already being considered and investigated in the early 1950s, but many years were to elapse before drifting was superseded in Britain by single and pair-ship mid-water trawling and purse-seine netting. These techniques were imported from Scandinavia and Europe during the latter half of the 1960s, by which time the development of horizontal echo-detection systems, designed *ab initio* to meet the needs of medium and small sized fishing vessels, had largely become the preserve of the Norwegian Simonsen Radio AS, better known as Simrad, which is today a world leader in the manufacture of both echosounders and sonars for fishing. Its echosounders are found in 20,000 vessels, including 150 research ships from many different countries. The early history of the commercial development of echosounders in Britain (Chapter 3), came about as the result of a close liaison between naval research departments such as the Admiralty Research Laboratory and the commercial companies Henry Hughes & Son and Marconi. The development of Simrad since World War II has followed a somewhat similar pattern. Furthermore, the company had the benefit from the outset of the expert knowledge and experience of a group of scientists who had spent the war years working with the British Admiralty Underwater Research Establishment at Fairlie in Ayrshire. That these opportunities have been exploited during the last 20 years with such outstanding success in a country which even today possesses a relatively limited electronics industry is a reflection on the size of Norway's fishing industry and of its importance in relation to the country's economy. This assured Simrad of a substantial home market for echosounders, as the value of this instrument had already been established by such pioneers in its use as Oscar Sund and Reinart Bokn before the war, and there was a ready-made demand greatly in excess of the supplies available from other countries at that time.

But the demand for sonar had to be created, or at least its value proved. Here Simrad had the advantage of an active and forward-looking Fisheries Research Unit at Bergen headed by Dr Finn Davold, as well as by the habits of the herring shoals themselves. The Fishery Research Vessel *G. O. Sars* (recently replaced by a second vessel of the same name) was fitted with a hybrid sonar, largely of British Naval design, for herring investigation. This was used most effectively for many years to find the large shoals of herring during the winter in the far north and to follow their movement southward some 100 miles off the coast of Norway to seaward of the deep trench that runs parallel to the shore. On reaching the latitude of Arlesund or thereabouts, the shoals would 'disappear', ie, they would go deep, below the beam of the *G. O. Sars*' sonar, and move inshore, passing under the south-going current that follows the direction of the Norwegian trench. The *G. O. Sars* would thereupon pass the news to the thousands of fishing vessels assembled at Arlesund and neighbouring ports waiting for the start of the 'great herring' season. Within 48hr these huge shoals would duly reappear close to the surface and swim towards the fjords to spawn, and the entire Norwegian fishing fleet would be waiting for them.

These invariably accurate forecasts depended on the magic of the ship's sonar, and so as soon as a set that was sufficiently cheap and compact to go into a 50ft boat was available, there was a ready market for it. The deep clear water off the Norwegian coast also facilitates horizontal echo detection, in sharp contrast to the relatively shallow and turbid waters off the east coast of Scotland and England. Moreover, the southerly migrations of herring on the British side of the North Sea have never followed so definite a pattern, nor have they been found in such dense concentrations; and because of the more difficult water conditions, the fisheries research vessels such as *Clupea*, *Clione* and *Ernest Holt* operating out of Lowestoft have not been able to rely in the same way on sonar for their detection. Abundance searches with echosounders, aimed at identifying the areas of greatest concentration during the season, have been a part of the Lowestoft Fisheries Laboratory's programme for many years, but the results of these endeavours have never entirely succeeded in securing the confidence of the fishermen. This is not to imply any criticism of the British effort, it is simply a fact of nature that the odds have been stacked against them. The fishermen want to know where the densest shoals will be tomorrow, or the day after, rather than where they are at the moment.

The timely appearance of the Norwegian 'great herring' shoals, hitherto so reliable, has completely changed within the last decade or so. Today the herring migrations are less regular than they used to be: they arrive later and move inshore as a rule much further north. A number of explanations have been put forward for these changing patterns but none have been proved conclusively. Over-fishing is often suggested as a reason. Another theory has to do with the cyclical changes of water temperature, which have a critical bearing on fish behaviour. Whatever the explanation, searching for fish shoals by sonar has become more than ever an essential part of the armoury of Scandinavian mid-water fishermen.

Although Kelvin Hughes took early steps to reinforce its substantial share of the echosounder market with a suitable sonar, it failed in the long run to match the local competition, and the great majority of sonars fitted in Norwegian fishing vessels are made by Simrad, with the British and German firms accounting between them for only a comparatively small proportion of the total.

Detection of fish, or for that matter of any other sort of object in the water, by the reflection of sound echoes is more difficult and more chancy in the horizontal direction than the vertical—owing to the nature of the sea itself and to the physical properties of sound. The principles involved have been described in Chapter 1. In one respect, however, where fish shoals are concerned, the advantage lies with horizontal detection, assuming of course that the water conditions are favourable. It often seems to be easier to assess the size of a shoal when looking at it horizontally than when passing over it. In the first place the angle subtended by the shoal, and consequently its size, can be quite accurately assessed by training the transducer across the bearing until the echo is lost at either edge. Taking the range of detection into account a rough estimate of the size can easily be made with a little experience. This technique also serves to eliminate at once echoes from other kinds of target, such as wakes, which cover a much wider arc and fade away at each end instead of disappearing suddenly between one transmission and the next. Furthermore, the length of the echo seems often to be significantly related to the thickness of the shoal as seen from the direction of the ship. Usually the trailing edge of the echo is as sharp as the leading edge, which is not the case when transmitting vertically. In this way the size of the shoal can be assessed with reasonable accuracy when lying off at a range of 500–1,000m.

In sets in which the transducer can be tilted vertically it is equally possible to measure the shoal's vertical height and depth below the surface. Even in the absence of this facility it is not too difficult to obtain a first approximation of depth by noting the range at which contact is lost during the approach. This occurs when the shoal passes below the horizontal sound beam, usually at a range equal to about six times the depth for a 10° beam. The exact depth can be confirmed with the echosounder when mid-water trawling, as the ship passes overhead, in time to set the height of the trawl appropriately.

To this day the types of sonar fitted in fishing vessels, though varying in many details of greater interest to the commercial fisherman than the general reader, are the same in principle as the standard types of sonar used for submarine detection during the latter part of World War II. They are generally referred to as searchlight sonars to distinguish them from the high-speed scanning systems which will begin to supersede them during the 1970s. Most searchlight sonars use an echosounder type of recorder, supplemented by a loudspeaker, the audio output obviating the need to watch the recorder chart all the time. Moreover the fish-shoal echo may often be heard before it can be seen on the chart, in contrast to good little Victorian children, who were traditionally seen and not heard.

To an experienced fisherman the echo is often recognisable and quite distinct at the outset from the many other objects, such as wakes, wrecks, rocks and ships, from which echoes may also be received. Even so, the first instinctive judgement should always be put to the test during the approach. Failure to do so has before now led an over-enthusiastic skipper to throw his net round a large boulder, with a consequent loss of both net and confidence. The various checks that can be made are, again, more a matter for the expert than the layman. The recorder itself is less than ideal for this purpose. When used for echo-sounding, for which it was originally designed, it produces a geographically meaningful picture for as long as it is switched on, ie, a profile of the seabed along the track of the ship. Used as the display for a searchlight sonar the past record has no value during the search period, and even after a target purporting to be a fish shoal has been picked up and held within the sound beam of the transducer, the trace gives only 'range-rate' information, as neither the manoeuvring of the ship nor the bearing of the transducer are recorded. But it still has the advantage of acting as a correlator, so that consistent echoes stand out against a random background.

While searching, the transducer is trained automatically, either in steps made immediately before each transmission or continuously at a rate appropriate to the range in use. There are several ways of doing this which affect the efficiency of the search from the point of view of the gaps left uncovered in the swept lane.

HIGH-SPEED SCANNING SYSTEMS

High-speed analogue scanning sonar systems were being developed experimentally in both Britain and America towards the end of World War II for naval use. In the fishing industry the principle has found its first commercial outlet so far only for research work. This decade will almost certainly see a more general application of the principle to fish detection and possibly for certain other purposes.

The speed of sound in water imposes a limit on the rate at which information can be obtained from any kind of single beam searchlight sonar. Increasing the range performance by using larger lower-frequency transducers is ultimately self-defeating, because the time taken to cover the whole area ahead of the ship is correspondingly greater and the gaps left in the swept lane become bigger. At the high-frequency short-range end of the scale the transmit-and-train method has an equally inhibiting effect on angular resolution, directly related to the minimum size of target that can be detected. For example, a 60° sector could be covered in this way by a high-frequency set to a range of 200m with a 10° beam in 1½sec, but to do so with, say, a 0·3° beam would take 50sec. High-speed scanning systems greatly improve the rate at which information can be obtained. In the example quoted the whole 60° is insonified by a single wide pulse, the 0·3° receiving beam is trained across the whole sector within the pulse time and the process is repeated continuously until the listening time representing the maximum range (¼sec) has elapsed. In this way it is possible to obtain the same amount of information in ¼sec that would otherwise have taken 50sec.

There is no difficulty in insonifying the whole search sector with a single pulse, although of course the wider beam means that the energy is more rapidly scattered. This can be compensated for up to a point by using higher transmitted powers. The essence of the problem lies in training the narrow receiving beam across the whole sector at the extremely high speeds required—far too rapidly to be done by mechanical means, but achievable electronically. The transducer is divided into a large number of small elements, each of which is separately connected to the main receiving amplifier. This enables the phases of the signals received in each separate section to be altered progressively across the whole face of the transducer in such a way that, when they are vectorially added together, the direction of maximum sensitivity of the receiver (the centre of the 'beam') instead of being at right-angles to the transducer face is shifted as required to either side. This is known as electronic beam steering.

Techniques have been developed by which this process can be carried out extremely rapidly, essential to enable the receiving beam to cross the whole sector being examined within the time of the transmitted pulse. Otherwise it would be possible for an echo to be received from one part of the sector when the receiver is 'looking' the other way, so leaving gaps in the search. There is thus a relation between the pulse length, the scanning speed and the receiver beam width. The narrower the beam and the shorter the pulse, the faster the beam must be steered.

To detect small objects in the water or on the seabed and to delineate the shapes of larger ones, range resolution, determined by the pulse length, is as important as angular resolution, determined by the receiving beam angle. The principle applies as much to long-range as to short-range systems. To take the example of a particular instrument developed by Dr G. M. Voglis and his colleague Dr J. C. Cooke at the Admiralty Research Laboratory (ARL) a good 15 years ago, the receiving beam is 0·3°, the pulse length 100μsec and the total sector angle 30°. In an alternative mode (which is made possible by the scanning method used as a kind of 'bonus') the same receiving beam can be swept across a 10° sector insonified by an even shorter pulse of 30μsec. In the former case the beam is steered across the sector 10,000 times per sec and in the latter 30,000 times per sec. When the narrower sector, therefore, is used, the pulse of sound in the water covers a length of under 2in and the target discrimination in bearing is of the same order, though increasing with range.

Scanning speeds of this order are realised by taking advantage of the fact that when two signals of different and varying amplitudes are mixed or 'modulated', new frequencies are generated.

These are known as modulation products, those of largest amplitude being the sum and difference frequencies, the so-called upper and lower side-bands. A change in phase of either of the two signals results in an equal phase shift of these side-band frequencies, either of which can be isolated by a suitable band pass filter. In the Voglis system the carrier frequency supply to each modulator is constant and differs by a fixed amount from those on either side. This difference equals the rate at which the beam is swept across the total sector. Thus, when the whole series of modulator frequencies changes from one unit in the array to the next by 10kHz, the sector is swept 10,000 times per sec. At the moments when all the modulator side-band outputs are in phase the whole transducer array will have maximum sensitivity to signals arriving from the centre bearing of the sector, ie, at right-angles to the transducer face. However, because of the differences in the modulator carrier frequencies, phase differences between the derived signals in adjacent elements of the transducer array will develop and increase to 180°, and then decrease until they are once more in phase at exactly the rate required to deflect the apparent receiving beam from one side of the sector to the other.

In other systems, such as the one developed much more recently by Prof D. G. Tucker and his associate, Dr V. G. Welsby, at Birmingham University, the carrier frequency is the same for all elements at any one moment, but is saw-tooth frequency modulated. This system involves the use of a delay line, and some distortion is unavoidable as a result of what can be called electromagnetic inertia. This is overcome by the Voglis method. For a full understanding of the techniques involved the reader should study the Bibliography at the end of the book (see Voglis, p 200).

The Voglis, ARL or bi-focal (so-called because of its two-mode operation) high-speed scanning sonar is therefore an extremely powerful short-range (because high-frequency) acoustic research tool, producing as it does a new picture displayed on a

'B' scan type of crt every $\frac{1}{4}$sec with a range and bearing discrimination unequalled even today. When it was first developed, it was outstandingly in advance of anything else of the kind.

For the most part, design details of within pulse sector scanning systems have not been made public until very recently. The design of a second generation prototype by Voglis and Cook at ARL has on the other hand been public knowledge for the better part of 10 years. It is now fitted in the MAFF fisheries research vessel MV *Clione*, where it has established itself during the last 2 years as a powerful tool for all kinds of research in marine science. This came about as a result of an investigation using this equipment into fish behaviour made jointly by the Ministry of Defence (MOD[N]) and MAFF in 1964. Details of the results were published subsequently in *Ultrasonics*. The pictures of fish shoals and trawls produced by this instrument were so outstanding that it was decided to hand it over on 'permanent loan' to the Fisheries Research Laboratory at Lowestoft, though the transfer took over 5 years to effect. The long delay was mainly due to troubles in designing a suitable stable platform, which arose out of a decision to add a new facility by which the sonar could be made to scan in the vertical as well as in the horizontal mode.

In 1964, the very same year as that of the MOD/MAFF trials using the Voglis system, the National Research Development Corporation (NRDC) made government money available to develop the Tucker system, which was then still in an early stage of development. No explanation for this decision, on the face of it a strange one, has ever been made public. The Voglis system had already been to sea in prototype form and, as the fishery trials conclusively proved, the concept was brilliantly successful. The Tucker system, by contrast, was still in an early stage of development and, in one particular at least, the principle used was technically inferior. It is surely inconceivable that NRDC were unaware of the Voglis prototype when they placed the contract.

Three or four years later the Voglis system was again considered for development into a production equipment, this time to meet a requirement of the Hydrographer to 'widen the furrow' in survey work. This is a somewhat different problem for which it is doubtful if high-speed scanning is the right solution, even in principle. The subject is considered more fully in Chapter 5.

Very little has ever been published on the non-development of the Voglis system, and the only reference to it to have appeared recently was contained in a report by the Natural Environment Research Council (NERC) Working Group on underwater acoustics, published in December 1971.[4] The relevant paragraphs are worth quoting:

> Development of a prototype analogue sonar for research purposes, based on the University of Birmingham patent, was jointly financed by NRDC and Coastal Radio Ltd. in 1964/65. The development was not completed by the firm, and the incomplete equipment was hired by DAFS and put into working order. It proved an unreliable construction but when working gave a good performance.
>
> The Working Group considered the problem of commercial manufacture of high grade sector scanning sonar and set up a small group in February 1969, to discuss this with NRDC. It was concluded that, of the two analogue prototypes available the ARL equipment should be redesigned for commercial development and there would be sufficient market from fisheries and research needs to warrant production. This recommendation was made to NRDC but there was no successful outcome. It was clearly recognised that there were other important applications of this instrument beyond fishing. The original equipment has been loaned by MOD(N) to MAFF, Lowestoft, where it has been installed complete with mechanical stabiliser on the mv *Clione* and undergone promising trials. A presentation was given at Lowestoft in November 1970 on the successful results ob-

> tained with this equipment, which MAFF now regard as a proved and powerful tool for fisheries research and marine science generally.
>
> In May 1969 a working party of the interdepartmental Government Committee on Marine Technology was set up to clarify the national requirements for sophisticated sonar equipment. . . . This Working Party considered the range of applications, but knowing of no exact specification in the fisheries field, concentrated its attentions initially on surveying and other applications. . . .

If NRDC, NERC, CMT and the rest were divisions of a private organisation, the shareholders would no doubt expect a fuller explanation of how their company had managed to dissipate a 10 year technical lead over its competitors, as well as a good deal of money in the process. As far as commercially profitable instruments for use in the fishing industry are concerned, Britain certainly has now lost the lead she once held in 'sophisticated sonars'. There are now three new sonars undergoing final proving trials at sea that will shortly be available on the market. All have been developed by industry with government backing. Two are within pulse scanning systems and the third is a multibeam device, and they come from Canada, Japan and Norway respectively.

The Canadian High Speed Scanning sonar, known as LSS-30, was designed by the Industrial Development Board of the Canadian Fisheries Service, and manufactured by C-Tech Ltd, of Ontario. In this set the transmitted pulse covers the whole 360°, which is swept by the 10° receiving beam once every 2ms, this being the length of pulse. It is designed to detect fish shoals in mid-water to a maximum range of about 6,000ft (1,800m) and uses a frequency of 30kHz, ie, ten times the range and one-tenth the frequency of the Voglis design. For searching at full range an alternative mode is provided by which the all round transmitting pulse is replaced by three directional beams 120°

apart, which each scan their own sectors within the pulse time of 10ms. This is done to reduce the spreading loss. The method introduces a small range error which increases across each sector to a maximum of about 20m, of no significance for fish detection. The information is displayed on a 10in crt in which the spot, brightened by the signals received, follows the path of a helix from centre to outside edge, synchronised with the rotation of the receiving beam, and complete once for each reception period. The result is a ppi display in which all targets are shown at their correct ranges and bearings relative to the ship, at the centre. The transducer array is cylindrical in shape, made up of a large number of separate elements, arranged in thirty-six vertical strips or 'staves', each containing twelve elements, a total of 432 in all. The receiving beam is rotated by electronic switching between staves at a rate of 555Hz, and an ingenious method of combining the outputs of each element is used to give a smoothing effect between adjacent channels to produce an uninterrupted display of the target. The same principles are being used for the design of a net sounder at a higher frequency with a transducer of correspondingly smaller dimensions.

A first production model underwent sea trials in Scottish waters in September 1972 in collaboration with the White Fish Authority (WFA). It was the subject of a paper at the Oceanology International 72 conference at Brighton, presented by the designer, L. W. Proctor.

The Japanese set is made by Furuno and appears, on paper, to have a less comprehensive specification. The scanned sector is limited to 120°, which may be centred on any desired bearing forward of the beam. Frequency, pulse lengths and range scales are similar. This set has not yet been seen in Britain.

A somewhat different approach has been adopted by Simrad. Using electronic beam steering techniques, a group of narrow pulses, each pointing in a different direction, are transmitted (frequency 38kHz) and received simultaneously. Ten beams are provided, each covering a sector of 6° × 6°, so that with each set of transmissions all echoes within a 60° sector are recorded in their correct positions, both as regards range and bearing. This is achieved by means of a flatfaced transducer divided into 121 square magnetostrictive nickel elements (see plate, p 34). The preformed beams are tapered in both planes to reduce side-lobe interference. Additionally, the transducer itself can be trained and tilted mechanically so that any 60° horizontal or vertical sector around the ship can be examined. The whole underwater unit is stabilised against roll and pitch and in azimuth. It is housed in a streamlined dome which allows for operation at acceptable noise levels up to a ship's speed of about 20 knots. This 'Multibeam Sonar' is part of a larger research project that includes a computer and a data display showing all the relevant information on a television type screen. The main market envisaged initially for this instrument is in the fishing industry as an aid to the skipper during the tricky process of ringing a mid-water shoal with a purse seine net. To this end a range of 500m has been selected as the correct one for a close contact sonar during the catching phase. When scanning in the vertical mode, it could also be used for seabed mapping, pipe-line surveys and in connection with dredging and port construction work. This instrument is not yet in production, but the prototype has for some months been undergoing extensive sea trials.

It is as much as anything the revolution in electronic technology that has brought these high-speed scanning and multibeam techniques to the point where they have become commercial propositions. The increasing use of integrated circuits in the next decade ought to bring down the prices of these complex miniaturised devices and ensure their reliability. The opportunities for development in this field are thus very promising.

It was the particular genius of the Voglis design to have achieved such remarkable results without benefit of the elegant building bricks available today.

At the research stage in Britain at various universities, government laboratories and a few private firms there are a number of projects that have potential for the fishing industry, though commercial production, if that stage is ever reached, lies still in most cases some way in the future. These include such possibilities as focused arrays, non-linear acoustics, the use of active and passive acoustic expendable radio buoys, sonars for airborne fish searches, alternative electronic methods of transducer stabilisation and vlf long-range detection systems.

There has been no reference in this chapter to American progress in the application of sonar to fish detection. Echosounding is, of course, widely used, but the catching methods, as well as the habits of the fish, in some important fisheries are not conducive to the use of sonar. The important menhaden fishing industry on the eastern seaboard is a case in point. This species of coarse fish, used for industrial products, shoals densely in shallow water near the surface; these shoals are best spotted from the air and both helicopters and aircraft are used for this purpose. An exception to this is the successful use of a continuous transmission frequency-modulated (CTFM) sonar for the study of schooling fish such as the bonito tuna by the Bureau of Commercial Fisheries. This type of sonar, developed by Straza Industries for search work in deep submersibles such as *Trieste I* and *II*, *Alvin* and *Deep Quest*, transmits a saw-tooth frequency-modulated pulse. Returning signals when mixed with the transmission produce a beat frequency which depends on the range. Investigations are also being made into the use of doppler signatures to evaluate the possibility of determining species and relative sizes.

Notes to this chapter are on page 194

5

SURVEY

HYDROGRAPHIC SURVEY

HYDROGRAPHIC SURVEY HAS A DISTINGUISHED HISTORY STARTING, in Britain, with the early navigators and explorers such as Cook and Flinders and continuing to the present day when the Admiralty Chart has a worldwide reputation and continues to be so called even after the Admiralty itself has become submerged in the Ministry of Defence. The first Hydrographer was appointed in 1795 and the present incumbent (1971) is the twentieth in line of succession, an average time in the chair of about 9 years, though 5 is the norm today. Perhaps the most famous was Rear-Admiral Sir Francis Beaufort who became Hydrographer in 1829 when he was 55, the age at which Hydrographers normally retire today, and held the post for 26 years. However, the French Naval Hydrographic service is the oldest, having been founded in 1720, and Denmark was next, in 1784. America followed, after Britain, in 1800, and the mid-nineteenth-century American Hydrographer, Matthew Fontaine Maury, is remembered for his pioneering work on ocean currents and for making the first mid-ocean bathymetric survey.

Until 1930 surveys were made with the aid of lead and line and the sounding machine. The first ultrasonic echosounder developed by ARL in 1930 was tested in a survey launch off Sheerness and in HMS *Flinders* soon afterwards. The Hydrographer of the day, Sir John Edgell, pronounced it an instant success. It was the start of a revolution in survey methods that is not yet at an end. Much the greater part of the British continental shelf is still covered only by surveys made with the hand lead.

Since World War II the survey echosounder has developed special characteristics of its own, though in principle it remains very similar to the basic navigational type. A high order of accuracy is attained, and an open scale ensures that the depth recorded on the chart can be read with equal or better discrimination. The accuracy is typically of the order of plus or minus 3in (7·5cm). A fix marker enables the surveyor to put a line across the chart by a pedal-operated switch to coincide with each geographical fix.

One of the problems of echosounding arises from the fact that the reflectivity of porous sediment varies with frequency. Thus in cases where the bottom consists of an initial layer of 'thick soup', gradually becoming more solid with depth until hard ground is eventually reached, the 'depth' recorded is less with a high-frequency machine (100kHz) than a low-frequency one (20kHz). Some dual-frequency echosounders are produced for this reason. A high-precision survey echosounder (Atlas-Deso 10) made by Atlas-Elektronik of Bremen, with a number of advanced features (including a digital readout), offers a choice of any two of the following frequencies—10, 30, 80, 100, 210 or 560kHz. The echoes from both are shown simultaneously on the same chart. The higher, connected to the grey line amplifier, makes a paler mark that is easily distinguishable from the lower-frequency bottom echo recorded in the standard way. Thus the upper limit of porous mud can be shown as well as the hard bottom beneath it (see plate, p 104). The 'depth' the surveyor is concerned with, as is the navigator, is the point at which the bottom material is sufficiently solid to cause damage to a ship should she touch it or pump it into her cooling system. The optimum sound frequency to select this level has never apparently been exactly defined. According to Dr O. Leenhardt of the Oceanographic Museum, Monaco, 80kHz would do it.

The time-honoured methods of working up a survey from the raw echosounding and positioning record to the 'fair sheet' are extravagant both in time and labour. Complete automation is now technically possible but not yet altogether economic; the process of change from manual to automatic methods has begun, but it will take a decade or more to complete. Here we are only concerned with the initial part of the process, but one that involves the most difficult step of all—the addition to the analogue echosounder chart of a digital record of soundings that can be applied to a punched tape or other data storage system.

The bottom is almost always the highest amplitude echo received, and it also has a measure of temporal stability in relation to the transmission. However uneven the seabed, the time relation of successive echoes will not change significantly when considered as a percentage of the total transmission interval. Signal selection of the bottom echo, therefore, can be achieved with a high degree of reliability by a combination of amplitude and time-gating circuits in the receiver. Alternatively logic circuits which compare each sounding with its predecessor may be used; this works well in shallow water, but a gating system may be more reliable in deep water when the signal-to-noise ratio is low. Heavy aeration can mask the pulse to such an extent that no bottom echo is recorded. If no signal that passes the electronic tests is received for several successive transmissions, the time gate is progressively widened and then opened completely, enabling the whole period between pulses to be searched. In such circumstances a high-amplitude echo from a fish shoal lying directly beneath the transducer might get the system off to a wrong start. Nothing could therefore be absolutely 100 per cent effective, but in practice such aberrations are exceedingly rare, and of little significance, provided the surveyor is aware of the possibility, so that a check survey can always be made in cases of doubt from the digital back to the analogue record.

The echosounder chart can act as a monitor by using the acceptance trigger to cut out the amplifier altogether for a short period. The gap produced in this way will become a white line following the contours of the bottom profile. Any interruption will indicate that the system has failed to select the correct echo at that point. Alternatively, the feedback monitor makes a mark above or below the seabed trace when the unit is digitising correctly. An incorrect depth may have been passed for digitisation but would not necessarily have been accepted. The prf of a shallow-water survey echosounder may be as high as 10 per sec, whereas one sounding every 5sec would be a sufficiently high rate of digitisation for most purposes. However, there are now systems which digitally record every sounding taken by these shallow-water sets. The mass of data resulting from this high data sampling rate—typically three-quarters of a million digits in a full working day—is, in 1972, proving easier in the collection than in the subsequent handling and processing.

When working from the analogue chart, the surveyor selects the least depth in each section, and in doing so takes out, by eye, the error due to the heave of the survey vessel. This is more of an art than an automatic process, and can only be attempted when the bottom itself is reasonably flat. The short-term oscillations rising periodically to a maximum that are superimposed on the record of the bottom profile can then be recognised for what they are. When the bottom is rough or stony or, worse still, in the presence of sand waves, the heave factor becomes more or less completely hidden, and accurate surveying from a boat is then only possible in relatively calm water. If a swell is present, the bottom profile will reflect this also by having a long, low, sinusoidal motion superimposed on it. This may be harder to detect. According to one authority, allowance made for the heave and swell factors results in an underassessment of depth by as much as 1m, but surveyors in general may dispute this.

There would be no great difficulty in designing the digitiser to scan the recorded depths since the previous digitised sounding and then select the shallowest. Automatic removal of heave and swell is a matter of great technical difficulty, and a completely satisfactory solution has not yet been achieved. There are two possible approaches: one way is to measure it directly by means of an accelerometer and the other is to fit a second high-frequency echosounder for measuring the doppler shift caused by the vertical motion of the survey vessel. By integration (double in the case of the accelerometer and single for the doppler method) the heave factor can be derived and added to, or subtracted from, the depth reading. Some experimental work has been done on both methods but neither has yet been perfected. Kelvin Hughes have experimented with the first method, using a Datawell accelerometer. The chief problem in using this principle that remains still to be solved is the elimination of a phase shift error, introduced by the accelerometer, which increases with the length of the heave period from 10° for a 4sec period to 90° for 33sec. However, until this correction can be automatically calculated and applied, by whatever means, some of the advantage of a digital read-out will be lost.

Each recorded bottom echo comes from the point at which the expanding sector representing the wavefront of the pulse of sound first comes in contact with the seabed. If this is flat, the echo will be from the point directly beneath the ship, but if there is a steep slope or ravine, the first echo may be from a position to one side, and the recorded depth will be less than it should be. These side echoes usually have characteristics that a surveyor can recognise on the chart.

In a survey carried out by C. G. McQ. Weeks, Chief Surveyor of Decca Survey Systems Inc, within the Kwajalein Atoll, in the Marshall Islands, there was a problem of distinguishing coral heads lying directly on the transit line from side echoes. The echosounder being used was the dual-frequency Atlas machine already referred to. A solution was found by reversing the conventional arrangement and showing the lower frequency (30kHz in this case) with the grey signal. Because of the wider beam angle the side echoes were picked out on the recorder with a grey mark, while the genuine pinnacles, recorded on the higher frequency, showed up as a darker trace (see plate, p 104). To design a digital signal selector to identify and eliminate side echoes automatically would call for some very sophisticated circuitry.

At present, therefore, the work of correcting the echosounder data for heave, swell, side error and tide, and of correlating the least mean depth with position, is not yet fully automated. Even when the technical problems are completely mastered, the cost is likely to be prohibitive.

Sidewall hovercraft, hydrofoils and helicopters all provide means of running transit lines at high speed but not without difficulty. Aeration will occur on the face of the transducer unless it is well sited and housed in a properly streamlined form. The transducer can be in a towed fish, provided that its depth in the water and position in relation to the towing vessel are accurately known. At these high speeds automatic recording of position will in any case become essential.

It has been assumed thus far that the echosounder is the only ultrasonic instrument required for hydrographic survey, but in fact it has never been more than barely adequate for the task on its own. Its great weakness is that it can only give information of the depth directly beneath the survey vessel. However closely the parallel transit lines are run, there always remains the risk of an undetected mound, rock, wreck or other obstruction lying between them. Until 1965 shipping routes were normally surveyed to a least safe depth of 20m, but this has since been raised to 31m, reflecting the greatly increased draughts of large tankers and bulk carriers when fully laden.

Methods currently in use involve the investigation of the areas between transit lines by means of searchlight and sidescan sonars. These instruments can confirm the presence of obstructions standing above the floor of the seabed between the lines of soundings, and in some cases identify them, but they cannot give exact information of the minimum depth of water above them; and so whenever the depth of the sea floor is shallow enough to make any such obstructions a potential hazard to navigation, they have to be investigated by other means, such as the use of a taut wire sweep towed between two vessels, an 'oropesa' sweep of the kind used for moored-mine clearance, or by the use of divers. These techniques are described in detail in Chapter 4 of Volume 2 of the *Admiralty Manual of Hydrographic Surveying*, which was rewritten and published separately in 1969. Here we are concerned with sonar search methods, and it is only necessary to say that the others are time-consuming and require a degree of skill and judgement that increases with the minimum depth to be declared clear of obstruction.

The types of trainable sonar used for hydrographic survey are in principle similar to those used for detecting submarines during World War II, and their evolution as well as their design and operating requirements have already been described. The surveyor, however, uses the set to look for and to identify just those objects which the naval and fishing operators are at pains to disregard. These may be confirmed as wrecks, pinnacles of rock or coral knobs in addition to the normally less well defined echoes returned by changes in the configuration of the seabed. In common with the other users he must be able to recognise, in order to ignore, all those echoes that are caused by wakes of other ships and by discontinuities or boundary layers in the water itself which occur at the interfaces between fresh and salt water or between two bodies of water of different temperatures. Freshwater springs, numerous in certain areas such as the Persian Gulf, can produce echoes as sharp as those from solid objects and can sometimes only be identified for what they are by sweeping through them with oropesa gear to establish the minimum safe depth, while sampling the water for comparison with the salinity of the surrounding sea. As the hydrographic surveyor is usually working in shallow water, often crowded with shipping and in the vicinity of river mouths and estuaries, unwanted echoes, of which these are some examples, are likely to be numerous and troublesome. They cannot be eliminated automatically because, although they may differ markedly from the wanted types in amplitude, extent and sharpness, they are nevertheless built of exactly the same 'bricks' as all other echoes, including the background of reverberation against which they are normally heard. They are the more troublesome for the fact that in the worst cases they are strong enough to present a barrier to the sound beam to the extent of masking echoes from concrete targets lying beyond them. In the presence of such conditions the surveyor will normally cover these areas from at least two directions, and perhaps from all four sides of a square. These precautions are necessary in any case where wrecks are prevalent.

Ideally a wreck will return a strong sharp echo of limited extent, easily distinguishable from other, woollier targets, but this is by no means always the case. When the wreck is in a tidal stream, it can be surrounded by eddies that effectively conceal its true nature; or it may have become partially submerged from one direction by a build-up of sand or unstable material on the seabed; from another direction it may, over the years, have caused a scouring out of the bottom so that the wreck echo itself may merge and tail off into an extended echo more typical of a sand dune than a solid object.

As has been explained in relation to A/S warfare, when searching with this type of sonar, gaps occur along the edges of the swept lane, caused by the fact that the forward travel of the ship is appreciable in relation to the time taken to complete a step-by-step search from one beam to the other. A mathematical treatment of the 'ideal' situation, ie, one based on the assumption that the search range selected is the same as the capability of the set in the prevailing conditions and that no time is lost in investigation of doubtful targets or areas missed through masking, proves that the optimum angle off the bow at which to start the search (on each side alternately) is independent both of step angle and search range. This is shown to be approximately 30° abaft the beam, allowing for a 'bow overlap' of 5° to cover the area near the ship's track lying inside the minimum detection range of the sonar. The best speed of advance is then shown to be a function of the ratio of maximum to minimum range.

The other type of sonar used for navigational surveys transmits pulses of sound on the beam bearing only that are narrow in the horizontal plane and wide in the vertical. Known as the sidescan or fixed beam sonar, it also has its origins in A/S warfare. The basic principle involved has already been described in Chapter 2. This instrument is a powerful tool for geological surveys, for which purpose it was first developed commercially, and for other specialised investigations involving seabed mapping. These applications will be discussed later. For hydrographic survey, ie, for surveys made for the specific purpose of preparing a navigational chart and for establishing minimum safe depths in navigational channels, its value is more limited.

The remarkable sidescan sonar trace of a wreck off Falmouth shown in the plate on p 137 is the negative of the original chart. The ship must be visualised as steaming up the left-hand edge, the distance between the vertical lines is 50ft, and the line separating black from white near the first interval mark, interrupted by the outline of the wreck and its shadow, represents the bottom profile. It will be seen that there is evidence of silting 'above' or beyond the wreck and of scouring on the near side. The first part of the chart, as far as the bottom profile, represents a vertical cut through the water—exactly the view that would be obtained with an echosounder—whereas the rest of the chart is a *horizontal*, or plan, view of the seabed. The transducer was in a towed fish, and as it passed directly over the fore part of the wreck, it must have come close to hitting it. The height of the part of the wreck directly beneath the survey ship can be estimated as being equal to the width of the shadow at that point (about 25ft), ie, halfway to the transducer which was being towed at 130ft in 180ft of water. The wreck appears to be on an even keel but in fact is not. If the mast were vertical, the sound beam would have crossed it in one or two transmissions and its shadow would have been horizontal. Since, in this case, a whole series of echoes all at the same range were returned by the mast, it must be canted diagonally away from the ship at an angle such that the slant range is constant. Knowing the course and speed of the ship, the heading of the wreck could be worked out and from this the angle of tilt could be derived. It will be seen that the shadow of the crosstrees continues off the end of the chart, showing that the line from the transducer to the top of the mast must be approaching the horizontal. The shadowed areas within the envelope of the hull are evidently caused by the upperworks, hatches and bridge superstructure. From a full analysis of all the information on this chart a pretty exact model could be made of this wreck in its final resting place at the bottom of the sea. But this example of the capabilities of the sidescan sonar is far from typical.

In practice, the information provided is not accurate enough to be relied upon. Silting, or scouring, on the far side of a wreck may cause the shadow to be, respectively, shorter or longer than it would be if the bottom were flat and level. Shadows are only cast by objects that are wide enough to interrupt a substantial portion, if not the whole, of the transmitted pulse as the ship passes by. If this condition is not met, some of the sound energy will reach the 'hidden' part of the sea floor beyond and return an echo from it which will mark the chart where the shadow should be. A mast or derrick, which could well be the highest point of a wreck, might thus be missed, a risk which clearly increases with range, however narrow the sound beam may be in the horizontal plane. The shadow limits are often blurred and the many different objects giving indications of one sort or another can confuse the picture to an extent that makes interpretation difficult or impossible. Moreover, the simple relation between length of shadow height assumes that the ray paths of the sound beam are straight lines, a condition that seldom occurs in practice. The sidescan sonar is thus a useful aid for examining the bottom between lines of soundings and for classifying the more prominent objects, but the information is qualitative rather than

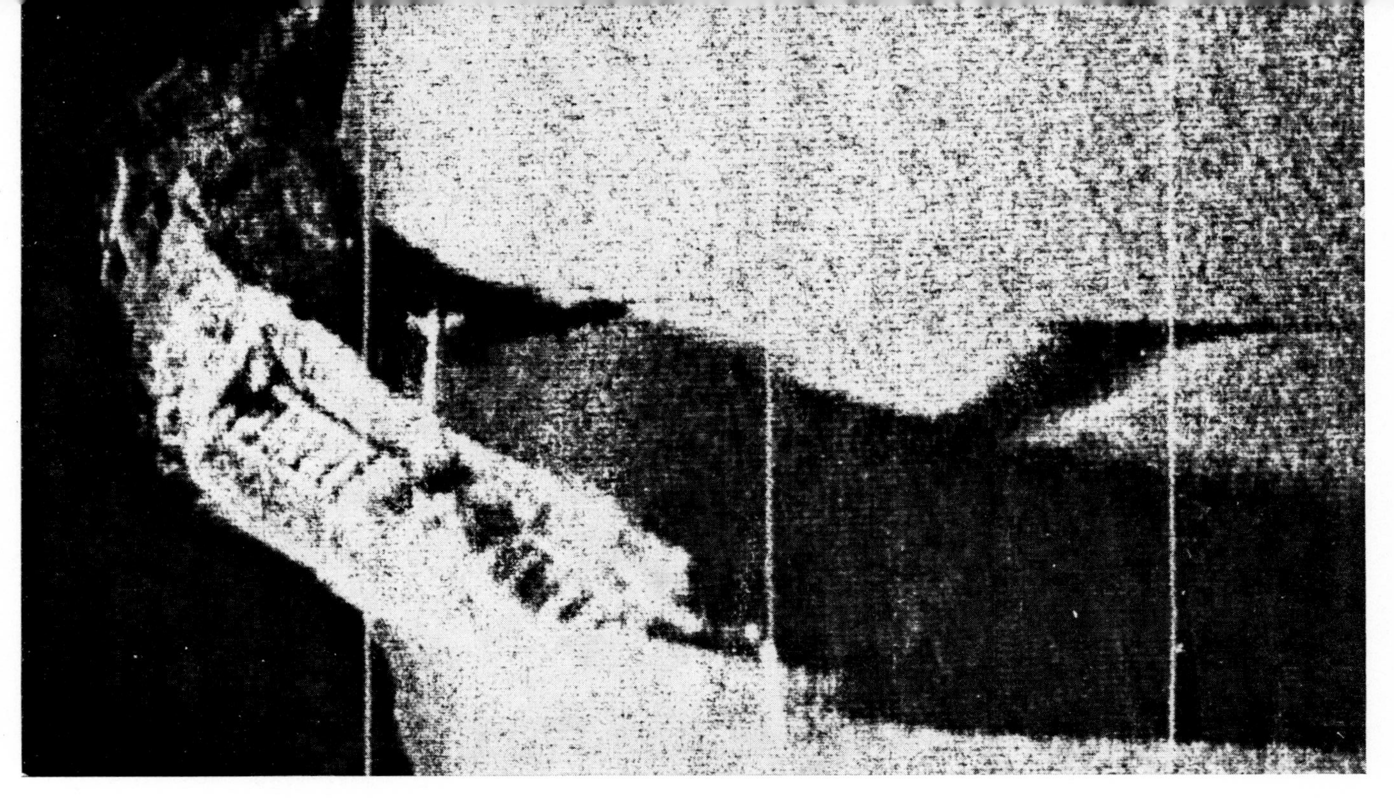

Page 137 *Sidescan wreck trace*. This is the negative of the actual recording. The vertical divisions are 50ft apart, approximately the height of the transducer (in a towed fish) above the seabed. Shadows within the forepart of the wreck indicate the hatches and show where they have been stove in. An analysis of this chart will be found on p 135

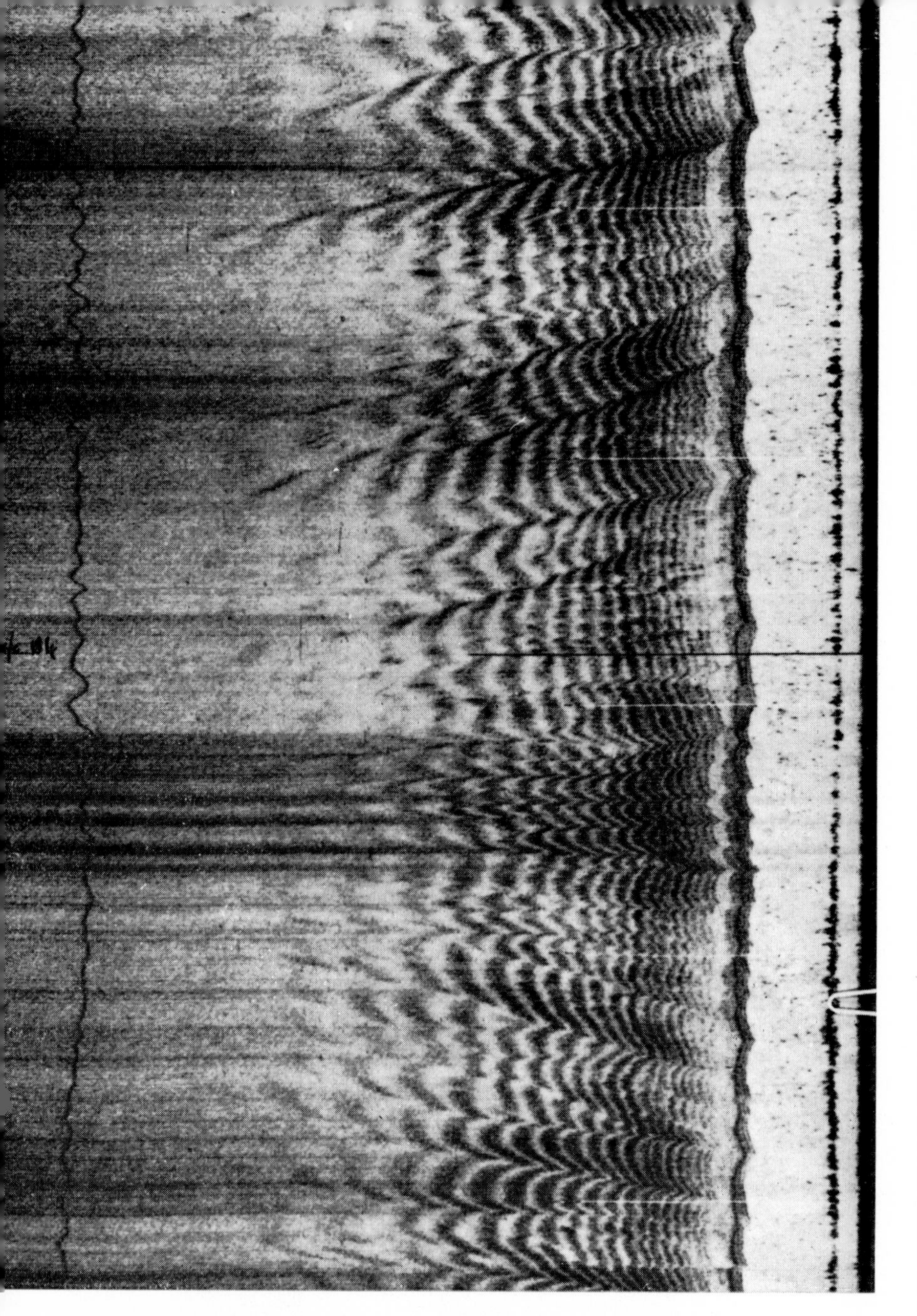

Page 138 *Telesounder chart*. The interference pattern is produced by using two transducers to transmit and receive the sidescan pulse. The thin white horizontal lines represent 1 minute intervals, and the black line half way up marks an alteration of course; this explains the apparent change in direction of the sand-dune ridges at that point. Sixteen interference lines out of a theoretical total of thirty-two can be identified. The mean depth is 40m and the wavy line up the left-hand side of the record indicates the heave of the vessel, obtained with a Datawell accelerometer

quantitative. It can provide valuable negative information by confirming the absence of obstructions that could represent a navigational hazard, but when the indications are positive, the surveyor still has to fall back on non-acoustic methods for exact measurement.

What is required is a new type of ultrasonic instrument designed to provide complete depth information over a lane of water covering, perhaps, 200m on either side of the survey ship —an instrument, as the Hydrographer has expressed it, 'to widen the furrow'.

There are some formidable technical difficulties in the way of achieving this target. Nevertheless, a successful solution is just about within reach of the present state of acoustic technology. The real problem is financial, since the cost of a system to meet the requirement in full would be prohibitive. The greater part of the work of the privately owned contract survey companies is not for navigational purposes. Their biggest group of customers are from the offshore oil industry who require specialised surveys of various kinds at all stages of the work of exploration and exploitation. The system for converting a line of soundings into a lane is not essential for any of this work, though would greatly facilitate the survey of estuaries, harbours, navigational channels and shallow coastal waters. This work is undertaken in the main by harbour authorities and civilian or naval hydrographic departments. The former, with possibly one or two exceptions, do not command the necessary financial resources, while the latter can argue that their primary function is to support the Armed Forces, notably the Navy, and that for this they do not need such equipment. The beneficiaries of deep-water hydrographic surveys, which need to be repeated at frequent intervals in areas where the seabed material is unstable, are the operators of deep-draught tankers and bulk carriers. These vessels are built because the resulting economy in manpower makes it profitable to do so. The financial calculations, however, do not take account of the real cost of surveying the shallow terminal waters of their routes, since no system has yet been devised to bring these costs to bear where they belong.

The Swedish Hydrographer introduced a system several years ago which can be considered as a step towards the 'lane objective', although unsatisfactory in some ways. This is the concept of a parent ship with a number of ancillary craft on either side, keeping station in line abreast, each with its own echosounder. The end result is a series of closely spaced profiles, accuracies of which depend on the station-keeping of the ancillary vessels. Furthermore, the movement of this convoy through a high-density shipping area would be quite impracticable. The Swedish lead was followed by the Hydrographers of Denmark and Germany, who developed echosounding sets in capsules towed from booms or carried in station-keeping floats somewhat on the principle used in oropesa minesweeping. These devices have some value in rivers and in calm water, but are basically unsatisfactory. The echosounders must be numerous and closely spaced if no gaps on the sea floor are to be left—which is the whole purpose of the idea—and any ship with boats in station or towing floats is unmanoeuvrable in confined channels.

The difficulties involved in producing a lane of soundings from one source spring from the fact that the accuracy of the result depends upon both the range *and* the angular discrimination of the beam, whereas the accuracy of the echosounder depends on the former only. Range discrimination is related to pulse length and sufficient accuracy is achieved when this is of the order of 40μsec. Bearing discrimination depends on directivity, and for comparable results the beam angle (to the half-power points) must be 0·1°, which involves a large transducer even at high frequencies. Moreover, when a flatfaced transducer or 'linear array' is used, the width of the beam can never be less than that of the transducer itself. Wavelength (and therefore frequency) and directivity are directly related to transducer size.

Assuming a frequency of 500kHz, corresponding to a wavelength of 0·3cm, a linear array to give a beam angle of 0·1°

would have to be 500λ, that is to say 150cm wide. The beam will diverge *from this minimum width* at an angle of 0·1°. The absorption loss increases with frequency and 500kHz is about as high as it would be safe to go without risk of losing the bottom echoes altogether from the outside edge of the lane. The target figure for accuracy suggested by the Hydrographer is plus or minus 6in (15cm) out to a maximum plan range of 200m either side of the ship. This cannot be achieved with a flatfaced transducer.

However, the required angular resolution can be reached by focusing the beam. Ways of doing this are being investigated, independently, by Kelvin Hughes and by the electronic department of the University of Birmingham. In the former system a group of small transducers are placed near the focal plane of a spherical reflector or mirror. The reflected beam of each transducer passes through the geometrical centre of the mirror, and reaches a partial focus at a range which depends on the distance between transducer and mirror. The diameter of the beam at the point of focus is considerably less than the width of a comparable flatfaced array. The hydrographic requirement could be met by a number of small transducers, appropriately positioned near the focal plane of a spherical reflector to produce a series of separately formed beams all looking out in the plane at right-angles to the fore and aft line of the ship but at varying angles of depression. Each beam would produce a profile of the sea floor at a different distance from the ship on either side. A visual display would not be too hard to design, and the bottom echoes from each beam could be converted to depth and digitised, subject to the limitations already discussed, for automatic reduction, correction and printing on to a fair chart.

A stabilisation system at least as precise (ie, to 0·1°) would be needed to maintain accuracy. The underwater acoustic reflectors would be much too large for an outboard mounting (but several smaller reflectors could be used in place of one big one) and would have to be built into the hull of the ship. Scale-model trials have proved the system in principle, but, though theoretically promising, it would be costly both to develop and produce.

Other practical solutions, which could not meet the standards of accuracy called for but might be better than nothing, may be considered. Techniques for 'anti-tapering' the directional beam pattern could be adapted to develop a beam having accentuated sidelobes that would look like the fingers of one's hand spread out and held sideways. This has already been done in some sort in an early design of geological sidescan sonar. Bottom echoes would be returned to the ship at discrete intervals from each 'finger' in turn, and as each would have a fixed angle of depression, the slant ranges could be converted into plan range and depth to give a series of closely run profiles which would, after digitisation and correction, be applied directly to the fair sheet.

A third method that has some possibilities, in theory at least, involves the application of interferometry to the sidescan sonar principle. The Lloyd mirror effect, the acoustic analogy of which was first noticed by Dr A. B. Wood over 50 years ago, has been the subject of a few studies in recent years, though some aspects of the phenomenon have never been fully investigated. At certain times, usually but not invariably when the surface is calm, the sidescan sonar map exhibits a superimposed interference pattern. This is caused by the existence of two paths by which the echo from the bottom can return to the ship—one direct and one reflected via the surface. When the difference in length between the two is a multiple of the wavelength, the echoes reaching the transducer reinforce one another. Halfway between these points the two signals are out of phase and there is cancellation, leaving a gap in the record. It will be seen from Fig 12 that these points of reinforcement lie at the intersections of a family of hyperbolae found in the vertical plane with the sea floor. If the bottom remains flat, the interference pattern will consist of a series of straight lines running vertically down the chart; if it starts to shoal, the lines will close up towards the transmission, and vice versa. Provided that the wavelength

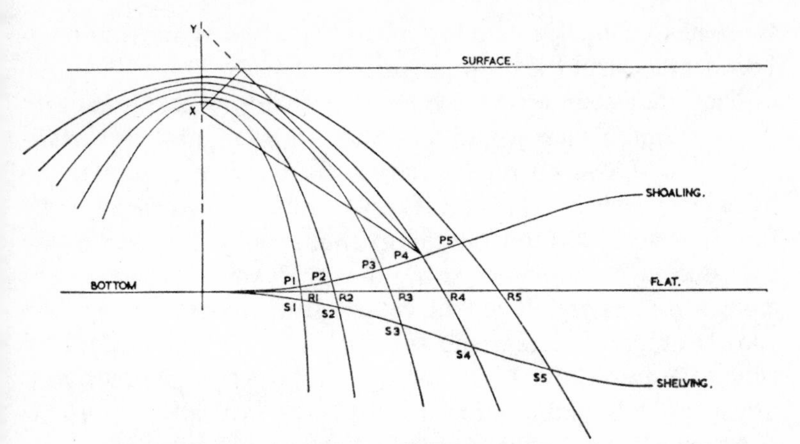

FIG 12 Lloyd mirror interference is liable to occur on sidescan records because reflections via the surface provide an alternative to the direct path between the seabed and the transducer. It will be evident from the diagram that equal path differences fall on a series of hyperbolae in the vertical plane. When the difference is equal to a multiple of the wavelength, the two signals reinforce each other and produce a stronger signal on the chart at the ranges at which they intersect the bottom. From these marks, striation lines are formed on the chart as the ship proceeds. The distances between these lines increase with range. When the ground is shoaling, they close up, and when it is shelving, they move further apart. Provided that the multiple number of any particular line is known (this is analagous to the problem of lane identification in hyperbolic fixing systems) the depth at each point of intersection can be calculated.

multiple can be identified for any particular line (analogous to the problem of 'lane identification' in hyperbolic radio position fixing systems), the depth of water at the point indicated on the chart can be calculated, and similarly for all the other lines that can be separately seen. They could be used to go a little way towards meeting the hydrographic requirement of running simultaneous parallel survey lines. There are some limitations to the usefulness of the principle: it is usually impossible to distinguish more than half the lines that are theoretically present, and those nearest the ship are lost when the intervals between them become less than the pulse length.

They are also interrupted by rock outcrops, wrecks and other obstructions, which would still have to be separately investigated. The Lloyd mirror interference is, itself, too uncertain to be of any practical use, but it is easy to envisage two *direct* paths for the returning (and outgoing) sound pulse by having two transducers, mounted one above the other, when the same principle would apply. It would be necessary to ensure that the Lloyd mirror interference never occurred, which can be done by tilting the transducer by 10° or so about the long axis. As with other methods, stabilisation against roll is critically important.

An extended investigation by the National Institute of Oceanography into what it has christened 'telesounding' (see plate on p 138) shows that depths measured in this way can be quoted to about 1ft. The ultimate accuracy will depend on a number of factors, of which stabilisation is the most important. Even with a computer the analysis of the data to yield vertical depths at the density required by the Hydrographer will be a tedious business, but no more so, and perhaps less, than from the equivalent number of echosounder transit lines—and, of course, the correlation will be much better. But the main advantage of the system is in gathering a lane of soundings with a single transit.

Almost 4 years have elapsed since the lane survey requirement was first raised, during which time the several government departments, working parties and committees charged with the study of this problem have been wearing their considering caps. They have consulted industry and several firms have put forward proposals for solution. All were rejected in favour of the Voglis high-speed scanning system on the grounds, among others, that this was the only one that had reached the prototype stage, which is somewhat ironical in view of the previous history of its non-development, as described in Chapter 4.

Moreover, the application of Voglis, or any other high-speed scanning system, to the survey requirement is open to some fairly fundamental objections. Such systems are designed to show every target within the searched area in its correct position relative to the transducer. This is obviously desirable when thinking in terms of a horizontal display either for examining objects on the bottom or in mid-water, but when such a system is turned on its side in order to examine a vertical cut through the water with the object of measuring minimum depths between sounding lines, it can be seen that high-speed scanning is a less attractive approach. For a start, the system provides too *much* information. In the Voglis arrangement, for example, a 30° sector is swept 10,000 times a second, whereas the surveyor is only looking for a density of echoes across his track comparable to that which he now gets along it from an echosounder. On the other hand, the angular resolution, fine though it is, would have to be improved by a factor of three to meet the accuracy requirement.

GEOLOGICAL SURVEY

The purposes of geological surveys vary from the needs of pure research into the structure, history and dynamics of the sea and the seabed, to the practicalities of the exploration and exploitation of offshore deposits of hydrocarbons, aggregates, minerals and so on. Although the presence of the sea imposes its own special problems for the geologist, it is not without its compensations. It has protected the seabed, particularly in the deep oceans, from erosion, so that characteristic structures are more easily identifiable. For seismic work the sea has the advantage of solving the difficult 'ground coupling' problem experienced on land, and provides mobility, so that a series of noise sources can be generated at intervals of a few seconds while the survey ship, towing both projector and hydrophone arrays and carrying all the necessary instrumentation, moves forward.

The limitations of the echosounder in profiling the sea floor increase with depth. Standard types of oceanographic echosounders were only capable, until recently, of delineating bottom features in the deep oceans of a kilometre or more in size. Smaller irregularities were smoothed out because of the large cross-section of the wavefront on reaching the seabed. When the bottom is sloping, an error is introduced in the depth recorded by a wide-beam echosounder, and it increases with depth. The angle of the slope is also incorrectly measured (Fig 13). Deep-sea photography, on the other hand, can only give data of objects measuring up to about 1m. However, the objects of greatest interest to the geologist are generally larger than 1m and smaller than 1km. Today they can be studied with the aid of narrow-beam stabilised echosounding systems. The following are examples of such instruments in service:

1. Narrow-beam transducer sounder (NBS) System made by the General Instrument Company of Massachusetts and used by the United States Naval Oceanographic Office (USNOO).
2. Narrow-beam echosounder (NBES) made by the Harris ASW Division of General Dynamics and installed in the US Coast & Geodetic Research Vessel *Discoverer*.
3. Narrow-beam sounder (NBS) made by Electroacustic Gmbh of Kiel and fitted in the DHI research vessel *Meteor II* (see plate on p 85).

Such instruments are costly because of the conflicting requirements of a low frequency to reduce the rate of attenuation and a directional beam to achieve good resolution of the seabed contours. To obtain these characteristics by conventional means, the transducer has to be very large. For example, a 2° beam to the half-power points at 12kHz requires a transducer face 3·7m in diameter. A way round this difficulty has been found within the last few years by the development of what is known as non-linear acoustics. This technique involves the propagation of two high frequencies simultaneously, both of large amplitude. Because of the slight non-linearity of the water, the two frequencies

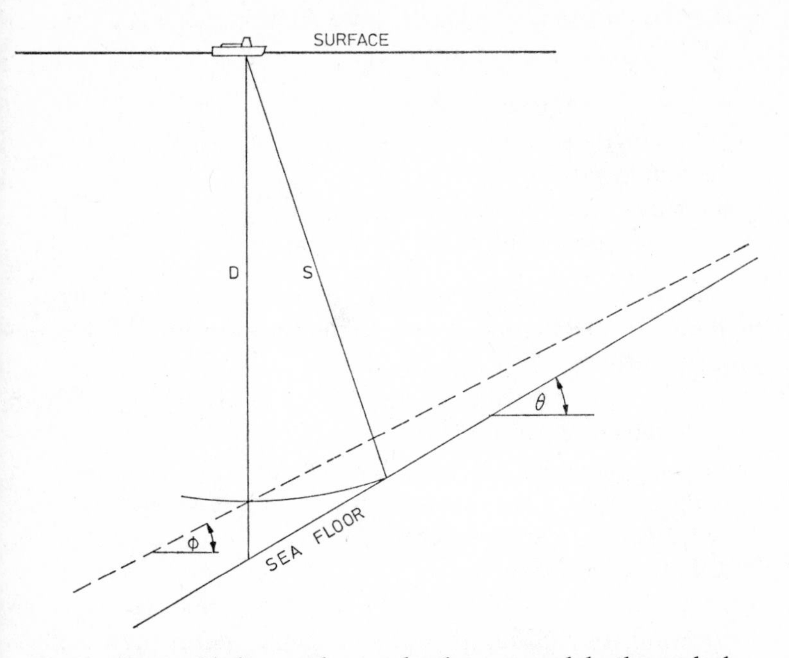

FIG 13 For a wide-beam echosounder the measured depth equals the true depth only when the sea floor is level, or when the ship is directly over a peak. At other times the measured depth is generally less than the true depth. In the case of a sloping floor both the depth and the slope will be incorrectly measured, as illustrated in the diagram. The true slope, θ, is related to the measured slope, ϕ, by the expression $\sin\theta = \tan\phi$

are mixed in the area in front of the transducer to produce a new acoustic signal at the difference frequency; this signal has the beam width of the high frequency components and the attenuation of the much lower, difference frequency. Moreover, side-lobes are absent from the directional beam pattern produced in this way. The process is inefficient, but this can be compensated for to some extent by increasing the transmitted power. Such a system developed by the Submarine Signal Division of the Raytheon Company Ocean Systems Centre at Portsmouth, Rhode Island, has produced some promising results on trial.

The basic parameters of their Finite Amplitude Depth Sounder (FADS) are as follows:

operating frequency—12kHz
pulse length—0·5ms
maximum prf—4/sec
transducer face diameter—0·225m (instead of 3·7m)
acoustic beam width—2·0° (nominal).

Profiles at a depth of 4,000m have been obtained so far by this method. The technique is also of importance for sub-bottom profiling, which will be discussed later.

An experimental study of the subject is also being made in Britain under a government grant at the University of Birmingham. Here the work has been directed towards the application of the principle for forward-search scanning sonar systems. It is hoped that this development may assist in the detection of bottom-feeding fish ahead of the ship, or to one side, an important and hitherto unsolved problem. Another possible application for non-linear acoustics would be in the development of doppler ground speed logs (Chapter 3), where a narrow low-frequency beam should help to increase the maximum depth at which speed over the ground can be recorded. The absence of sidelobes from the beat-frequency beam would be of particular importance in this case.

For deep-sea charting, the oceanographic echosounder, however directional it may be and however sophisticated the design, still suffers from the inherent limitations already discussed in connection with hydrographic surveying. It gives no information to either side of the ship's track. A new bathymetric charting system, described as a multi-beam array sonar, was recently developed and built by the General Instrument Corporation of Massachusetts for use by USNOO. This is a complex system in which a series of projectors mounted parallel to the keel (from thirty to sixty, depending on requirements) transmit a fan of sound over a 90° arc at right-angles to the ship's track. Each projector is individually pulse-shifted to compensate for pitch. The receiving hydrophones are mounted athwartships to produce a crossed-fan system giving a narrow-angle beam in both planes from each part of the fan. The preformed beams are individually processed electronically and stabilised for roll, so that each of them represents a side-looking angle independent of the ship's roll and pitch. The coverage of the seabed is directly proportional to the depth of water, being about 2 nautical miles (3·7km) at a depth of 1,850 fathoms (3·4km). The signals are digitised and fed to a computer, which calculates the reference position for each bottom point in relation to the ship and the vertical depth at that position. This information is fed to an X-Y plotter on which the cross-course contours are automatically drawn in real time at the intervals and scales appropriate to the depth and terrain. An example of a contour strip about 2 nautical miles wide is shown in Fig 14.

Used in shallow water the coverage of this particular system would be insignificant, certainly nowhere near the 200m either side of the ship that the Hydrographer has asked for. Probably the accuracy would be inadequate also for navigational survey.

It is in the work of the marine geologist and geophysicist that the sidescan sonar has proved its value, even more perhaps than as an aid to hydrographic survey. For seabed mapping, initially within the continental shelf areas and more recently in the deep oceans, the instrument has been used on its own and in conjunction with sub-bottom penetration sounders and seismic instruments (which will be described later) as well as non-acoustic devices such as magnetometers, underwater cameras and coring and bottom-sampling equipment. Over the years much general experience has been gained in the difficult art of interpreting these sonar records which appear in the most fascinating variety of shapes and patterns, some almost photographic in their detail, others overladen with all manner of interference patterns. Two examples are illustrated on p 155. While other methods, such as those referred to, are often essential for

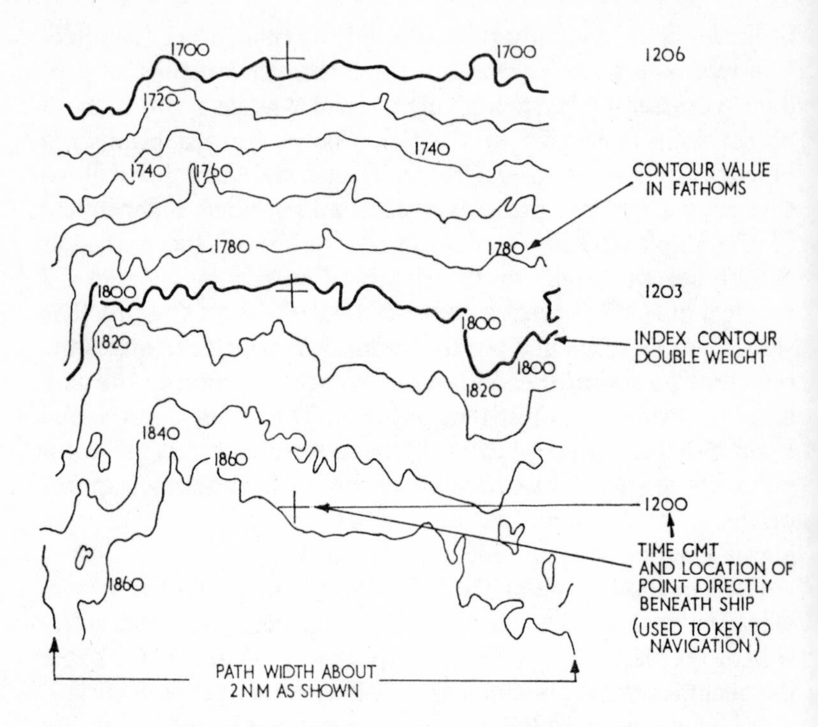

FIG 14 An example of a contour strip as derived automatically from the United States Oceanographic Office Multi-beam array sonar

positive identification of the features displayed acoustically, the sidescan sonar is nevertheless unrivalled in its capacity to produce a coherent map consisting of a mosaic made up of long strips hundreds of yards in width.

It is as well to remember, however, that even the most apparently straightforward and evocative 'pictures' of the sea floor produced in this way can be misleading. The sonar display is obtained by 'indirect' methods, after all. The way in which an acoustic shadow beyond an object on the seabed, such as a wreck, can delineate its features, has already been explained; but shadows, or what appear to be shadows, can also be formed in other ways. A trench, or shelving bottom, or any area hidden from the direct line of sight from the transducer will appear white. Such areas may extend beyond the edge of the chart, but if they do not, a darker mark will often be seen on the further side of the shadow, instead of the nearer. In general the smaller the angle between the sound beam and the bottom, the smaller will be the part of the echo that is back-scattered to the transducer. However, this effect is compensated for somewhat by the law of the time varied gain (TVG) circuit in the receiving amplifier, which is designed to allow for the fact that as the range increases, the returning signal becomes weaker due to attenuation and also to the increasingly acute angle at which the sound beam strikes the bottom. Ideally, in a perfectly set up machine, a flat bottom of unchanging material will produce a signal of constant strength right across the chart.

The amount of back-scattering also varies according to the nature of the bottom. In practice no material is smooth enough to reflect the sound as a mirror, so that there is always some scattering of the signal, the amount depending on the grain size of the material deposited on the sea floor. Therefore, a distinction can be made between mud, sand and gravel.

In addition to the complexities of the seabed map itself, further confusion may be caused by a variety of effects that are not directly related to the nature of the bottom. The fan-shaped beam may pass through mid-water targets such as fish shoals, wakes and so on, the echoes from which will be superimposed on the basic pattern. Propeller and other machine-made noise can appear in various guises. From the ship itself this may show as a pattern of alternating dark and light areas, not unlike the typical Lloyd mirror interference, but in this case the spacing of the striations will not vary across the width of the paper. Noise may also be encountered from many sources outside the ship, such as other ships' propellers and the quenching effects caused by bubble sweep-down, especially in rough weather. In tidal waters the movement of the sea across ridges and sand waves, as well as the transport of unstable material on the sea floor, causes noise over a wide spectrum of frequencies. Experience with the Voglis high-speed scanning system shows that this is still prominent at 300kHz. The sounds made by marine mammals, crustacea and to a lesser extent some species of fish may also be seen. Such forms of 'passive' interference do not so much confuse the record as obscure it, causing heavy horizontal bands across the chart whenever the sound beam is in line with the source.

Another disturbing factor associated with the sidescan system when used for seabed mapping is distortion. This is caused in several ways. The first and most fundamental is due to the non-linearity of the display.

Fig 15 shows that the distance of any echo mark from the leading edge of the chart (representing the moment of the

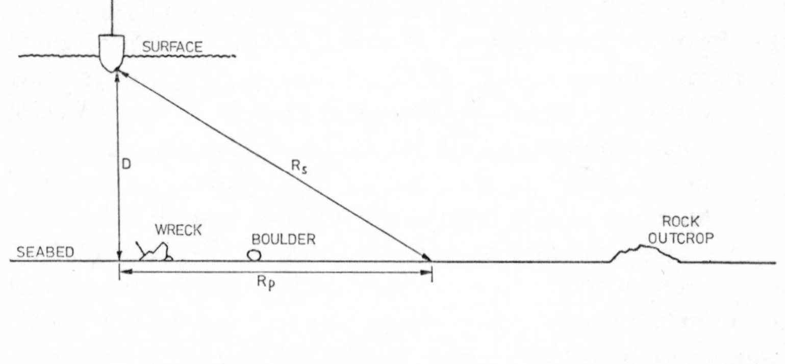

FIG 15 The sidescan sonar measures the slant range, R_s, whereas it is the plan range, R_p, that is required for seabed mapping. The error, R_s–R_p, is a function of depth and decreases with range. In the diagram the sidescan sonar trace is shown below the drawing to the same scale. It will be seen that the record is considerably compressed near the start, so that the wreck and boulder are shown too close to each other, whereas the rock outcrop, at a greater distance from the ship, is only slightly out of position

transmission) is a measure of the slant range (Rs) as opposed to the plan range (Rp). This means that the seabed map is compressed to the maximum extent at the point where it begins, directly beneath the ship. At maximum range the two dimensions Rs and Rp approach equality. Being a function of depth it would be difficult to allow for this distortion automatically on the recorder display and, in fact, there is no recorder on the market that does this. A more important form of distortion, from the point of view of the geophysicist, is the difference between the vertical and horizontal scales of the chart record. The relation between these two scales depends on the one hand on the range in use and the width of the chart, and on the other upon the paper speed and the speed of the ship. Some of these factors are interdependent. For example, the paper speed is related to the prf (and therefore the range), as trace-to-trace correlation is necessary to obtain a coherent picture. The correct paper speed for any given prf is much higher for a moist paper recorder than for a dry one because of the difference between the two stylus widths.

Normally the vertical scale, ie, the one in the direction of the ship's travel, is the more compressed, and the remedy is either to reduce the range scale of the chart or the speed of the ship. Neither solution is altogether satisfactory.

Alternatively, the chart can be 'rectified' subsequently by optical means. The French company Geotechnip does this optically by means of a group of spherical lenses mounted between two groups of cylindrical ones which can rotate in opposite directions round the overall lens axis to achieve the appropriate anamorphosis.

Another way of producing an isometric display has recently been developed at Bath University. By this method the signals are recorded as an intensity-modulated display on a crt that is then photographed one frame at a time on 35mm film. This gives the required flexibility in the vertical time base, which can be set at the correct value to yield an isometric picture. The Bath method can also correct the distortion caused when the survey ship is steaming across a tideway.

In common with other acoustic instruments a compromise has to be made between the conflicting requirements of range and resolution, and there is today a wide range of frequencies used in sets that are commercially available. These vary between 30kHz and 250kHz. Here are some examples:

Geomechanique (SOL–101–S)—37·5kHz
Kelvin Hughes (Transit Sonar)—48kHz
Klein Assoc Inc (SA–350)—100kHz
EG & G—105kHz
Westinghouse—150–160kHz
Thompson CSF (TSM5610)—200kHz.

Finally, in a class of its own, there is the long-range sonar known as GLORIA, developed by NIO, which operates on 6·5kHz.

When a picture is generated on both sides of the ship at the same time, it is possible for echoes from one side to be picked up in the opposite transducer, but these can usually be recognised by the absence of a shadow. Most sets on the market today use a towed fish for housing the transducers. This is essential for use in deep water and has the advantage of providing quieter operating conditions for the transducer; roll and pitch are greatly reduced if not completely eliminated and the decoupling effect of the towing cable also reduces the heave of the ship in rough weather which has a more disturbing effect on the picture than roll.

The use of this instrument for seabed mapping started in earnest 10–15 years ago, and since then many of the world's continental shelf areas have been mapped. The National Institute of Oceanography in Britain was the pioneer in this development. Its first design operated at a frequency of 36kHz, and the transducer was a magnetostrictive type made up of nickel stampings mounted in a framework outside the hull of the old

Page 155 *Sidescan sonar underwater maps*. (*right*) Examples of sand waves in the vicinity of the North Edinburgh Channel in the Thames Estuary. Some smaller waves, at right-angles to the main sandbanks, can be seen. (*below*) A double sidescan record of a sea floor strewn with boulders. The ship should be visualised as steaming up the centre of the chart

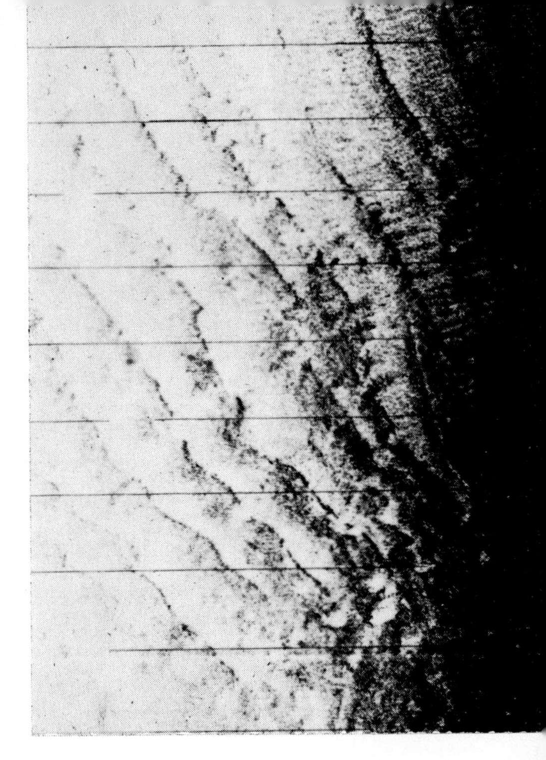

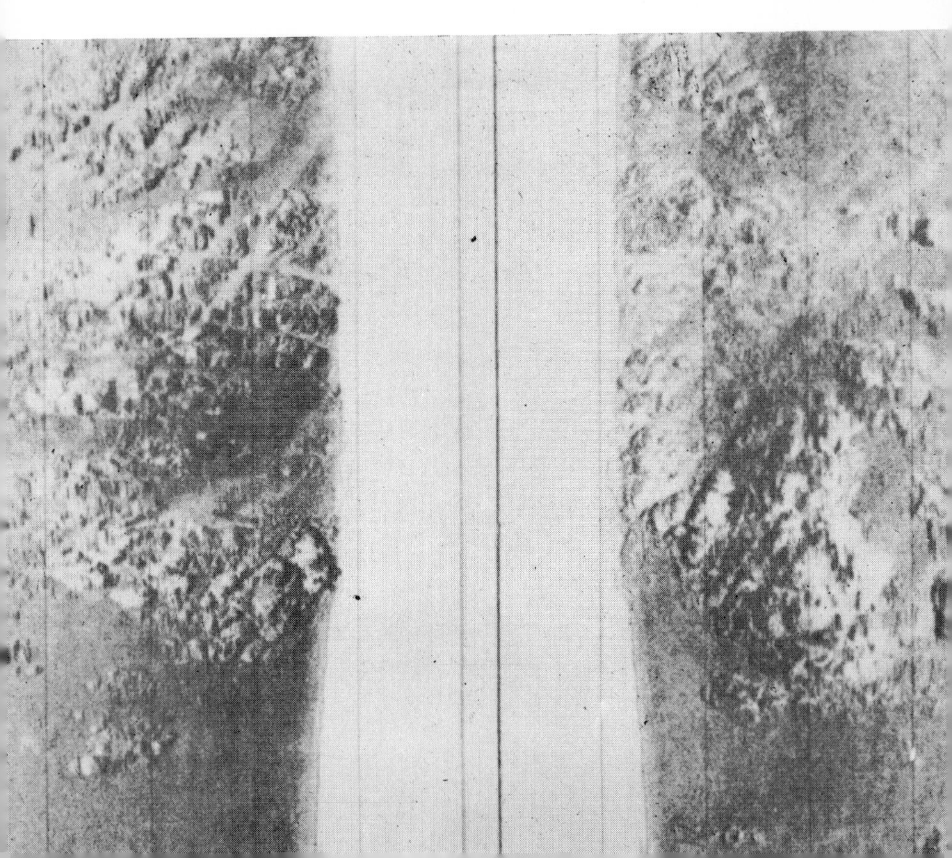

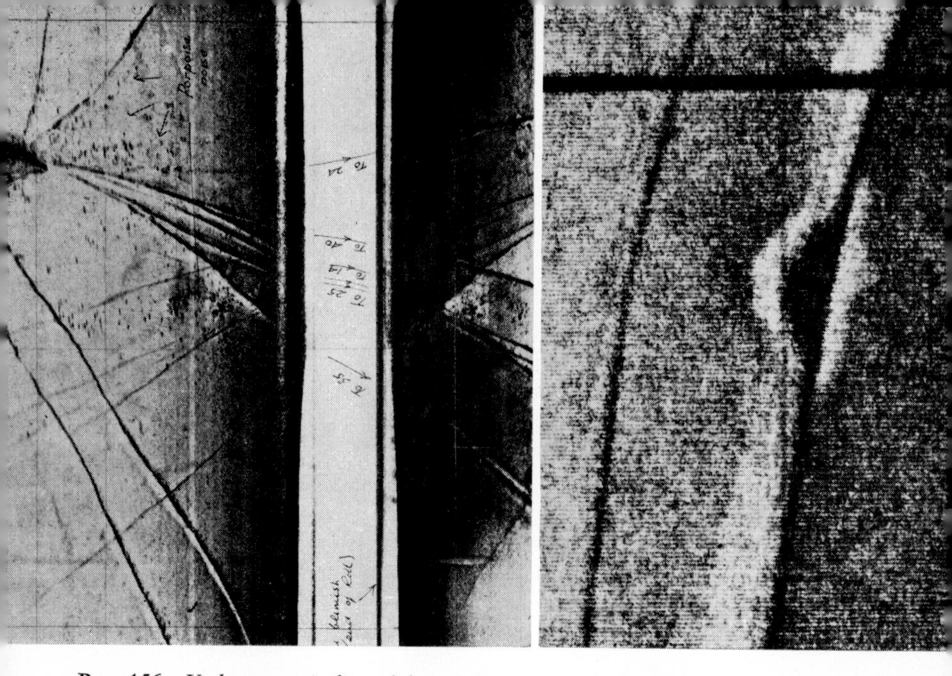

Page 156 *Underwater pipelines.* (*above left*) A network of pipelines, some on the seabed an some partially buried, emanating from a flow station in the North Sea. (*above right*) A magnifie section of a sidescan chart showing a section of pipe clear of the seabed. This is indicated by th acoustic shadow on both sides and by the thickening of the trace. The second pipe withou a shadow is a reflected echo trace from the other side (not shown). (*below left*) A low-frequenc echosounding record of two pipelines. The stronger echoes returned from the pipes than th seabed allow the former to be picked up by the sidelobes of the beam before the ship is over head, and held after it has passed. This causes the distinctive approach curve hyperbolas

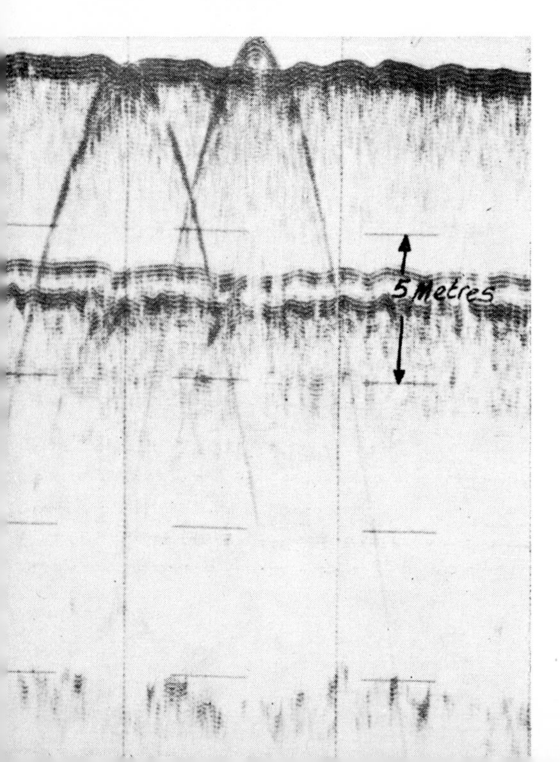

Discovery II. This has been replaced in the NIO's present research vessel, *Discovery*, by a piezoelectric type using barium titanate and operating at the same frequency. It is also stabilised against roll and can be adjusted from inboard to any angle between horizontal and vertical to give the optimum coverage for different depths of water. Facilities are also provided for increasing the amplitude of the sidelobes in relation to the main beam in the vertical plane. The effect of this is to divide the section of the record nearest to the ship into two or more broad bands separated by white gaps. These bands follow the contours of the seabed profile, though to an exaggerated extent because the sound paths are slanted. An outcrop of rock will show up clearly and the angle at which it lies to the ship's track will be made evident in this way. On the other hand, objects of interest can be missed between the lines.

A derivative of this design was used by BP for experimental work in the Persian Gulf and elsewhere, and a later version was used effectively and extensively by Prof Chesterman in the early 1960s at Hong Kong University. His investigations covered much of the harbour itself and its approaches, and later surveys were extended to the Macclesfield Bank in the South China Sea. This early work by NIO and at the University of Hong Kong led to the discovery of many new features and paved the way to an understanding of the mechanism of sediment transport in tidal waters. It also displayed the versatility of this type of instrument, which is now, in its many different forms, being used in all parts of the world. The *International Hydrographic Review*, *Acoustica* and *Nature* have all carried articles detailing this work.

In contrast to the trend elsewhere, which has been towards the use of higher frequencies to improve the resolution while accepting some reduction in range performance, NIO in 1964 decided to go low. The result of this decision eventually took the water under the name of Long Range Sidescan Sonar, or Geological Long Range Inclined Asdic, leading to the acronym GLORIA by which it is generally known.

The basis of the GLORIA project was to scale up the geometry of the typical sidescan sonar by a factor of 25 to give a theoretical maximum range of 12 miles (22km) and to be capable of bottom mapping beyond the limits of the continental shelf in depths up to 18,000ft. The design performance has been reached but only at the expense of considerable complication. The transducer array, 15ft (4·5m) in length to produce a 2° beam at 6·5kHz, is made up of 144 separate elements of a new design of piezoelectric lead zirconate titanate having a conversion efficiency of over 90 per cent. The towed fish, containing ballasting equipment, anti-yaw gyro, monitors for depth, roll, pitch, yaw and heave, and much else besides is 10m long, 1·5m in diameter and weighs 6·7 tons in air and 3·5 tons in water. It is towed at a normal depth of 400ft (120m) by a 130 core double armoured cable attached to a nylon rope accumulator system at speeds up to 7½ knots. On one occasion during the proving trials it was dropped and damaged: *sic transit* (very nearly) *GLORIA mundi*. The increased pressure at the prescribed towing depth has the advantage of permitting a higher level of transmitted power before the onset of cavitation. The recorder is a helical type using moist paper. There are two helixes, the second of which provides a picture compressed in range to the extent necessary to give an isometric view for a variety of ship's speeds and sonar ranges.

On completion of a series of handling, towing and calibration trials in Loch Fyne in June 1969 GLORIA was taken to the Mediterranean to test the strength of its arm in mapping and identifying various geological targets in deep water. The results prove that the instrument can detect targets at its maximum designed range, no mean feat in itself. However, these long ranges are only achieved at the expense of discrimination. The transmitted pulse extends for 150ft (45m) in the water and at 12 miles the beam is something like 2,250ft (650m) wide. Even so, some provisional interpretations of geological formations have been made. A survey south of Crete showed trends on the highly fractured 'cobblestone' region of the Mediterranean Ridge. In the course of another survey a fault, 400km in length, running east from the Azores, was located; this is believed to be the boundary between the Eurasian and African plates in that region. In 1971 GLORIA was used for a short time off the Western Isles of Scotland to carry out an experimental long-range herring abundance survey. Shoals were detected under favourable conditions at 10 miles (18km) and a purse seiner was homed on to the target. Three catches were made, with a landed weight of 240 cran (50 tonnes), the largest of the night.

Recent development has been concentrated on improving the quality and gain of the correlation circuits, and the quality of the display. A linear correlator with a 40dB dynamic range is now used, and the record is made on photographic paper by means of a modulated light source. The power gain achieved by the processor has opened the way to the design of a smaller, more easily handled sonar operating at a higher frequency but with a range comparable to that of the original design, which could well have wider survey applications.

In discussing the sidescan sonar in relation to geological survey, attention has been concentrated so far on its application to research work rather than to the more practical requirements of contract survey teams working directly for the offshore extractive industries. The main difference in these two kinds of activity with regard to design is that the second group must be cost-effective to survive. Not for them the costly one-off brain children of the research institutes, they must make do with instruments in production, transportable from one survey vessel to another and suitable for use in chartered ships.

Sidescan sonar has an important role to play in the inspection of pipelines already laid on the sea floor (see plate, p 156). When this is composed of shifting deposits, the pipe can start a process of scouring that may eventually leave sections of it hanging in mid-water, a potentially dangerous situation, as the extra strain could damage and eventually break the pipe. It is to

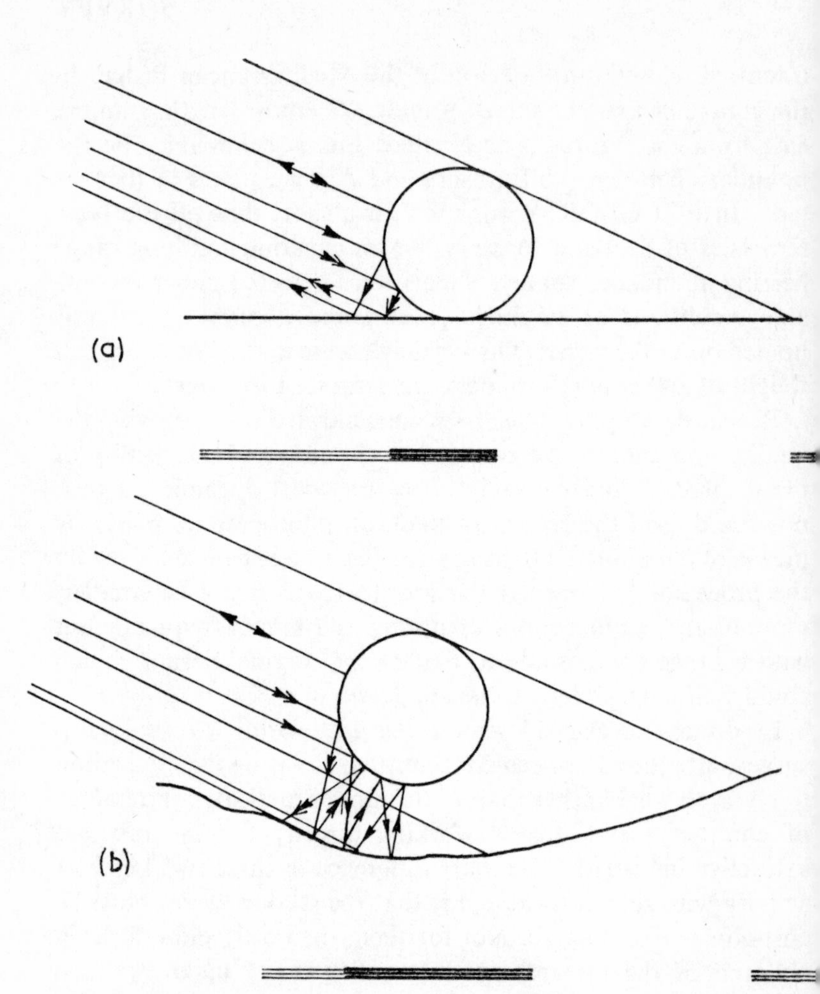

FIG 16 The two cross-sectional diagrams show, (a), a pipeline lying normally on a flat bottom, and, (b), a pipeline suspended in mid-water as a result of scouring. In each case the sidescan traces are shown to the same scale underneath. In (b) the darker part of the trace, representing the near side of the pipeline, is considerably extended into the shadow zone area. This is caused by the multiple reflection paths between the pipeline and the seabed which spread out and delay the returning echo. Some possible sound ray paths are indicated

be hoped that such a catastrophe will be avoided, for the resulting pollution would be on a massive scale. If it were a gas pipe, it has been estimated that the reduction in density of the seawater locally, caused by the escaping gas, would actually be sufficient to sink ships in the vicinity. When the pipe is lying clear of the seabed, the echo returned from it will be thicker than it normally is, extending out behind it into, and perhaps beyond, the acoustic shadow. This is because the part of the sound beam that strikes the lower half of the pipe is reflected downwards into the hole underneath it, and from there back to the pipe again and thence to the ship. Various paths for the sound are possible and these multiple reflections delay the returning signal so that the echo is extended (see Fig 16). The plate on p 156 gives an example of a trace obtained from a partially suspended pipe.

SUB-BOTTOM PENETRATION

The seabed acts as a reflector and a scatterer of sound in a complex way, depending on its nature, and when it is composed of sediments such as mud, sand or gravel, on its porosity. The acoustic impedance of this material, as compared with that of seawater, determines what proportion of the incident wave is reflected (and scattered) and what proportion penetrates the interface. As the sea floor is typically made up of sedimentary layers of varying characteristics and of different rock structures, the sound beam is split into reflected and transmitted parts at each boundary until the energy is dissipated. The reflected waves from the second and subsequent interfaces undergo the same process of splitting on the way back to the sea. The directions of the sound path are also changed in accordance with Snell's law as they cross each boundary. Fig 17 is a simplified diagram of what happens in practice. A pulse of sound that penetrates the seabed, to a greater or lesser extent, is thus split into a number of different paths, and these return to the hydro-

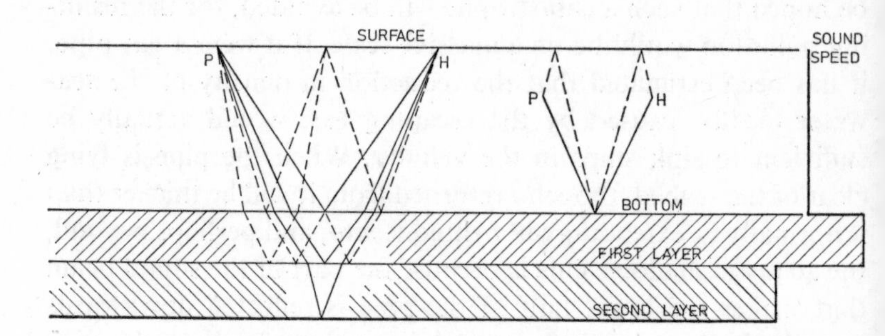

FIG 17 The diagram shows, in simplified form, some of the possible paths by which the sound can travel from the projector (P) to the hydrophone (H) in seismic surveying. Single reflection paths are shown by continuous lines, and multiple reflection paths, via the bottom, sub-bottom layers and the surface, are shown as dotted lines. For simplicity the sound speed in each medium is assumed to be constant. Its value is indicated on the right, on a relative basis

phone array in the form of an extended series of discrete echoes. In the early days of seismic prospecting it was assumed that each observed wavefront had been reflected only once. However, within the last 20 years or so it has become increasingly evident that multiple reflections, both via the surface and within sub-bottom layers, are seldom if ever absent (Fig 17). The identification of these multiples adds considerably to the difficulty of correct interpretation of seismic records. One further fact must be taken into account in the analysis: whenever the sound wave is moving from a lower to a higher velocity medium, the reflected wave undergoes a change of phase or polarity, and this fact, incidentally, makes it possible to tell whether some distant explosion is of natural origin or man-made.

For seabed penetration two types of sounds are used—single-frequency pulses, generated by transducers such as are used for other acoustic instruments, and noise sources of various kinds associated with separate receiving hydrophone arrays. The former are generally cheaper and more convenient to operate and can provide better resolution, but only limited penetration at best, and none at all in some cases. The latter, by releasing far greater energy, can penetrate the sea floor to a depth of several thousand feet but at the expense of resolution. The two systems have complementary uses, and each has its own special problems. Those using pure tones at the resonant frequency of the transducer are variously called mud probers, sub-bottom profilers or strata recorders. They give the best results in mud, clay and chalk sediments, with penetrations up to about 70ft (21m) in the right conditions; but are less successful when the overlying material is sand or gravel, when the critical factor is the diameter of the average grain, measured in wavelengths. If this is more than about 0·03λ, the energy is quickly scattered by contact with the individual particles. For this reason the wavelength must be as long as possible, and also, in order to reduce attenuation to the minimum, low frequencies are essential for this work. A frequency as high as 18kHz is used by ELAC of Kiel with some success but the most favoured range is from 3–12kHz. One of the earliest instruments was the Sonoprobe with a frequency of 3·6kHz. Good results were obtained with it by the Scripps Institution when pinging through chalk in an English Channel survey made in 1958. More recent work has proved the importance of a high transducer conversion efficiency, which is needed to generate the maximum possible acoustic energy. For this reason, in the second generation systems now in use, there is a trend away from magnetostrictive (nickel) and piezoelectric (ADP) transducers towards electrostrictive types using barium titanate and lead zirconate titanate. For certain investigations, such as the determination of the depth of gravel deposits and the examination of buried pipelines, a clear view of the first few feet below the sea floor is of primary importance.

A Raytheon development known as a Correlation Echo-sounder Processor (CESP) shows considerable promise in resolving this particular difficulty. In the CESP process a large bandwidth/long duration signal is transmitted and at the same time stored as a reference replica for returning echoes. A DELTIC Correlator is used to transform the relatively long wide-band signals into short pulses equivalent to the reciprocal of the bandwidth. This technique makes maximum use of the high resolution inherently available in the large signal bandwidth, and at the same time permits the transmission of a considerably higher total energy (power × time) than would otherwise be possible before the onset of cavitation. Background noise is also substantially reduced, since this does not correlate with the particular characteristics of the stored replica of the transmitted pulse. Experimental results show marked improvements over those achieved by conventional means.

When working in very shallow water, the first reflected echo via the surface, which is often extended by the immediately preceding echo returned via the bottom of the survey vessel, can be troublesome in masking the record just where sub-bottom reflections may occur. Altering the depth of the transducer is of no help, since the displacement of both echoes will be the same.

Probably the optimum results for all types of sediment will not be found at any one frequency, and the ideal arrangement would be a set with a variable frequency, such as the Thompson CSF type TSM5210/1, which gives the operator a choice of frequencies from 2–7kHz and pulse lengths from 100μsec to 10ms. Among other manufacturers not already mentioned who have designed bottom-penetrating sounders are EG & G (6kHz), EDO Western (7·3, 10·12kHz), ORE (3·5, 5, 7kHz), Alcatel (3·3, 5·9kHz), CGG (5kHz) and CESCO (1·5, 3, 6kHz).

In terms of depth, seismic technology begins where sub-bottom profiling ends. On the other hand, marine seismic methods have developed from those used on land, whereas sub-bottom profiling has followed the development of echosounding. During the last 15 years rapid strides have been made in marine seismology and the refraction technique inherited from prospecting methods used on land has now been largely superseded. This method involves the use of deeply buried explosive charges that set up oscillations which travel at different speeds, according to the densities of the underlying strata. An analysis of these trains of vibrations, as recorded on a seismograph at some considerable distance, enables inferences to be drawn about the thickness and composition of these strata. By repeating the process at discrete intervals a map of a vertical section through the earth's crust can be constructed. The process is slow, laborious and costly. Greatly improved methods of electronic position-fixing enable precisely orientated seismic surveys to be undertaken anywhere on the continental shelves and, beyond, across the deep oceans. Deep penetration requires brute force as well as low frequency and the fundamental problem is to put as much energy into the water as possible while at the same time avoiding the series of bubble pulses caused by the contraction and explosive re-expansion of the gas-filled cavity that is the inevitable aftermath of a deep-water explosive charge. This 'signature' cannot easily be coded or altered, and for this reason, as well as the fact that explosive sources cannot be used for continuous profiling, their use is now largely restricted to special air-dropped applications, where mobility and depth flexibility are of paramount importance. The methods commonly in use today for continuous seismic profiling consist of high-energy electrical discharge systems (Sparkarray and Hydrosonde), electromechanically vibrated plates (boomers and thumpers) and air guns in which air pressure is built up between shots and suddenly released.

In a recent paper given by Dr G. A. Gauss of Hunting Geology & Geophysics Ltd, the author gave the following figures for the performance of the sources commonly used by his company (see table overleaf).

Source	Frequency (Hz)	Penetration	Resolution (in water)
Sparkarray (Geodyne)	80–200	Up to several thousand feet	Worse than 15ft
Hydrosonde (Huntec)	500–800	Up to 100m/sec in rock	2ms or better (approx 9ft)
Boomers	500–800	Better than 100 m/sec in rock	2ms plus
Precision Boomers (EG & G)	1,000–1,500	Up to 50m/sec in uncemented strata	Better than 1ms

During the early 1970s marine seismic profiles were recorded, both for geophysical research and for prospecting, at a rapidly increasing rate. A new seismic prospecting ship recently delivered to the German company PRAKLA actually cost less than its equipment, which included satellite navigation, interfaced with a doppler sonar, and electromagnetic logs; various Decca parabolic position-fixing systems, with associated track plotter, data logger and onboard computers; and equipment for magnetometer and gravimetric survey. The sound source used is the compressed air gun, of which up to thirty are available in groups of ten. The streamer carrying 1,500 hydrophones is 1½ miles long. Base lines of this length are used in deep seismic surveying to separate out multiple reflections; the depth of the streamer is maintained at a constant level by active self-regulating buoys and the end one is fitted with automatic direction-finding equipment to maintain the whole streamer on its correct line. The signals received are applied via time-sharing cables to a bank of amplifiers and subsequently to an analogue/digital converter for storage on magnetic tape. A continuous record is also provided for onboard inspection. For deep refraction seismic work, for which the ship is also equipped, the hydrophone assembly may be anything from 3 to 30 miles astern of the explosive sound source and its position is then controlled by telemeter.

A large and growing number of oceanographic survey vessels are covering thousands of miles of seismic profiling every year, and before long there will be hardly an ocean basin or a continental shelf anywhere in the world that has not been traversed and mapped. A few examples will suffice to indicate the scale on which these operations are now being carried out. In 1969 the US Naval Oceanographic Office (Navoceano) and the US Geological Survey (USGS) joined forces to operate a year-long geophysical research programme in the Gulf of Mexico, using the Navy's Oceanographic vessel USNS *Kane*. The first phase involved a series of criss-crossing cruises over the deep basin in the western half of the Gulf. Deep seismic profiles were run in depths up to 8,000ft (2·4km) with sub-bottom penetrations ranging from 3·8 to 4·7sec, ie, 3–4 miles below the seabed. Teledyne Exploration equipment was used, comprising four 'sparkers' fired simultaneously at 4–6sec intervals to generate, together, 160,000 joules (W × sec) while maintaining a ship's speed of 9–10 knots. In shallower water this deep seismic equipment was used in conjunction with a 3·5kHz strata recorder to cover the first 200ft (60m) or so below the sea floor. The project led to the discovery of a large number of submarine domes beneath the seabed, predominantly in the Sigsbee Deep. There is strong evidence to support the view that these are salt domes, a fact which, if confirmed, greatly enhances the possibility of massive oilfields in the region.

Also in 1969 the Teledyne Exploration Company's own geophysical exploration vessel *Stranger* returned to base at San Diego on completion of a cruise that had circumnavigated the globe, and during which 4 million 'shots' covering thousands of miles of continuous seismic profiles had been recorded. In this case also a quadruple sparkarray was used, towed 100ft astern (30m). The receiving array, consisting of over 100 crystal hydrophones in a single streamer, was 500ft (150m) behind the projector.

Meanwhile, in the North Sea alone, well over £10 million has been spent on seismic profiling within the last 5–10 years. Two British Government-financed organisations in particular—the Institute of Geological Sciences (IGS) and the National Institute of Oceanology (NIO)—have been active in seismic and sub-bottom profiling research in many areas of the British continental shelf, using respectively RSS *John Murray* and *Discovery*. The latter has also been active in deep-water investigations in the Bay of Biscay, the Indian Ocean, the Red Sea, the western Mediterranean, and the north-east Atlantic, where a new seismic technique developed at NIO in association with Cambridge University, was tried in 1968. To get away from the relatively high level of ambient noise near the surface the instrument was to be placed on the seabed. Buoyant spheres of high tensile aluminium alloy, which can withstand high hydrostatic pressure, are used as instrument containers. The hydrophone signals are tape-recorded, synchronised by a high-precision crystal clock. The spheres are released from the seabed acoustically, and a tracking system, for use while surfacing, is supplemented by a radio transmitter and flashing light for surface location. These spheres have been successfully used in depths up to 2,600 fathoms (4·8km).

In addition to the much publicised offshore searches for hydrocarbons, seismic surveys have been used to look for other commercial treasure in many areas, including the following:

Gravel—UK and other continental shelves
Tin—off Malaysia and Thailand
Diamonds—off South Africa
Phosphates and other heavy metals—off Australia, South America, India, USA
Gold—off Alaska
Base metals such as iron—many inland lakes.

But it is in the search for new offshore oil and gas fields that the most profitable business for this type of survey lies. So it is the deep end of the boom industry that is really booming. The deep-shot seismic records provide inferences on which wild-cat drilling decisions involving millions of pounds are made. America, Canada, France and Germany are the maritime countries with the largest stake in seismic prospecting, but Britain, sad to say, with the exception of some research projects of the kind already referred to, is only in at the shallow end.

SIDESCAN AND BOTTOM PENETRATION COMBINED

The sidescan sonar and sub-bottom (seismic) profiling systems have been described separately in relation to their application to geophysical research and prospecting. Today these two types of instrument are being increasingly used in combination for many other purposes of a practical nature. These include preparatory surveys prior to dredging and pile-driving for harbour development, and investigation of the most favourable routes for underwater pipelines as well as post-lay inspections. These techniques are used also to find the most suitable sites within selected areas for the operation of 'jack-up' and 'semi-submersible' offshore drilling platforms. For this purpose it is necessary not only to choose a flat piece of ground free of rock outcrops and other surface obstructions but also to ensure that the material immediately below the sea floor is more or less of a uniform consistency so that the rig, when lowered into position, will settle evenly. A limestone ridge, for example, coming within a few feet of the sea floor but not outcropping would not be seen by sidescanning alone.

Similar considerations apply to the selection of a deep-water approach channel to a new oil terminal. When dredging is going to be required to achieve the necessary depths, a study of the topography in advance will enable a route to be chosen that involves the least amount of rock blasting. This may not necessarily be the shortest way. To simplify the record analysis in such cases it is an advantage to obtain the sidescan and subbottom records simultaneously and to the same scale. Several companies have developed combined systems, notably EG & G, Huntec and IFP.

SEDIMENT ANALYSIS BY ACOUSTIC METHODS

When the sound velocity of the sediment exceeds that of seawater, it can be measured by transmitting ultrasonic pulses from a probe driven into the bottom and moving a hydrophone by measured amounts away from the transmitter along the sea floor. The elapsed time computed for a series of readings gives an accurate figure for the velocity. In a similar way the attenuation can also be measured. With this information the material can be analysed on the basis of the following relations, which have been established in recent years. The velocity of sound in a saturated sediment is found to be closely related to porosity which, in turn, has a linear relation with the bulk density. The velocity of sound is also related to grain size and depends on size distribution. The mean velocity can thus be equated to sediment porosity, mean diameter and particle size spread about the mean. The attenuation coefficient is also related to the average grain size. In clays and silt it is found to increase with diameter up to a maximum of 60 microns and then to decrease. The size spread is also a factor. Two other parameters can be measured on board by means of special equipment to analyse the sonar echo returned from the sea floor. These are the reflectivity coefficient, which depends on the acoustic impedance and is also related to porosity, and the back-scattering density, which increases with grain size.

A large-scale investigation of the reflectivity of the floor of the north Atlantic Ocean was undertaken a few years ago by the British and Dutch Hydrographic Departments acting in concert.

The operation, named 'Navado', was carried out by the British survey ships HMS *Vidal* and HMS *Dalrymple* and the Royal Netherlands Navy vessel *Snellius*, which covered between them over 60,000 miles in seventeen transits from 10° N to the Arctic Circle. The study also included an investigation of the deep scattering layers and other sources of sonar interference. Detailed results have not been published.

6

COMMUNICATION AND CONTROL

SINCE THE LATE 1950S ACOUSTIC METHODS HAVE BEEN INCREASingly used in subsea information retrieval and remote control of operations, and in underwater communications. Purely ultrasonic methods have started to replace moorings in deep water for the steady control of platforms and drillships in precise locations over the seabed during drilling operations. It is probably in these areas that the most significant advances will be made during the coming years.

ACOUSTIC TELEMETRY

The acoustic telemeter is the alternative to an electric cable, still the more widely used method of passing information at a distance. Many activities in the sea will always require the use of divers. In waters that are too deep for diving the more complex operations are likely to be handled from submersibles with remotely operated tools. To such methods sound is not so much an alternative as a complementary technique. In recent years there has been a huge increase in the number, accuracy and reliability of data sensors available to the oceanographer, and the very immensity of the 'vasty deep' will ensure that the demand continues for many years to come. In the past the usual procedure has been to store the data in the collecting sensor unit for later recovery and processing.

However, it is often desirable and sometimes essential to retrieve and display such information in real time—for instance, in the control and monitoring of pelagic nets and other towed instruments. To obtain a sample from a deep scattering layer seen onboard in an oceanographic echosounder, the marine biologist requires a means of monitoring the height of his net. For the same reason a fisherman using a mid-water trawl must have accurate and up-to-date knowledge of its height above the bottom and also, ideally, a view of the shoal in relation to the net mouth as the trawl sweeps through it. Coring and dredging operators have similar requirements.

The first acoustic telemeters developed in the 1950s were of the single information channel type, suitable for transmitting slowly changing data, such as temperature or depth. The data is transmitted as a function of the time interval between a starting pulse and the data signal. This is normally followed by a blanking period when the receiver is blocked until the multipath effects have died away to negligible levels. These effects are caused by signals following other than the direct path, such as reflection via the seabed, the surface or scattering layers, and in other ways that are still not fully understood. Thus at the receiving end signals arrive in a confused and extended pattern, some in phase and some out of phase with the direct signal. Developments in acoustic telemetry, which are still continuing, have the objectives of increasing range and reliability and, above all, the information rate. Pulse modulation and pulse code modulation techniques, and ways of transmitting information from several sensors via a single telemeter, have been developed and are now commonly used. The latter either use a frequency-modulated carrier or a time division multiplexing system. Some telemeters, such as the series developed by NIO, are used also in a reverse direction, for passing command signals to the remote unit. As the transmitter must have an independent power source, the most efficient use of the available energy is of special significance. Various types of long-life compact batteries are normally used, but in the long term, fuel cells and power from the energy conversion of radioactive isotopes offer alternatives, of which the latter is probably the more promising in this application. The conversion efficiency of atomic energy is still, however, very low, and the units themselves are at present too expensive for general use.

One of the more important applications of acoustic telemetry is to pass information from the bottom or mid-water trawl to the fishing vessel. While trawling, the skipper needs to know the headline height above the bottom (or depth below the surface). He is also interested in such parameters as the width of the mouth and its aspect relative to the ship, the temperatures of the water, and the quantity of fish in the 'cod end'. A trawl telemeter has been developed and tried at sea during the last several years by the White Fish Authority (WFA) Industrial Development Unit. It is designed to transmit up to four sets of data from bottom trawls used for fishing in depths up to 540m (300 fathoms).

The four parameters measured by this system are:

Water temperature
headline height (0–20ft or 0–100ft)
depth of transmitter
attitude of transmitter in pitch plane.

Additionally, later development is designed to measure the width of the trawl mouth and the amount of fish in the net.

As the rate of change of these variables is low, multipath effects can be eliminated by blocking the receiver immediately after the receipt of each direct path signal. Changes in the position of the trawl relative to the ship, as well as yawing in rough weather, make it necessary to provide an automatic tracking facility to lock the receiver to the bearing of the transmitter at all times. This is done by measuring the phase gradient of the incoming wave in relation to the receiver array and then applying the appropriate phase shifts in the signal processing network. Time division multiplexing is used, and the information is conveyed by prf modulation of the pulsed carrier wave (55kHz to avoid echosounder interference).

Another acoustic telemeter developed specifically for the real-time measurement of the characteristics of fishing trawls, but of somewhat more recent origin, is the sixteen-channel digital system designed at the Marine Laboratory, Aberdeen. This model differs also from the WFA version in that it is designed primarily as a research tool rather than for use in commercial fishing operations. Its maximum range and depth performance is somewhat lower (1·5km and 400m respectively). It takes full advantage of modern developments in integrated and miniaturised circuitry and, being digital in concept virtually throughout, achieves a high degree of accuracy. Up to twelve parameters consisting of any combination of tensile loads, vertical heights, speed and depth can be measured and transmitted sequentially and, in addition, up to four distance parameters can be measured acoustically. Time division multiplexing is used, each measurement transmitted being preceded by a positive identification code. The measurements are passed by a form of binary code in which 'zero' is represented by a short pulse and 'one' by a pulse three times as long. Three decades are available for each measurement, giving an accuracy in transmission of 0·1 per cent. The receiving hydrophone is mounted in a towed body and fairly heavy towing cable serves to keep it well below the slip-stream.

The Ministry of Agriculture, Fisheries and Food (MAFF) Marine Laboratory at Lowestoft uses telemeters for similar purposes and has experience of both frequency and pulse rate modulation techniques. The National Institute of Oceanography (NIO) has likewise developed its own range of telemeters, tailored to meet its particular needs. These operate with a carrier frequency of 10kHz, much below the others already referred to, and are mutually compatible. Ranges of 3,000m have been achieved.

In America, Benthos Inc of Massachusetts has developed a deep-water acoustic telemeter specifically for oceanographic use which operates on essentially similar lines. The pulse frequency in this example is 12kHz and the length 2ms. A narrow-bandwidth receiver helps to reduce ambient noise. The timing pulse occurs at 1sec intervals, corresponding to the 400 fathom scale available in most oceanographic echosounders, with all of which this telemeter can be operated. It has provided precise depth information during benthic and pelagic trawling operations carried out by the Woods Hole Oceanographic Institution (WHOI). With this instrument Dr Ross of WHOI discovered some dramatic temperature changes—as much as 12° C in a few metres in one case—associated with the hot brine pools in the Red Sea. On other occasions it has been used to transmit heat-probe data and core alignment in conjunction with a WHOI deep-coring project.

The Benthos telemeter has also been used in conjunction with special monitoring equipment developed at the Atlantic Oceanographic Laboratory, Canada, to observe the performance in deep water of the Bedford Institute rock core deep-sea drill.

Inevitably, because that is where the money is, the most advanced underwater acoustic telemetry and command systems have been developed to control and monitor the complex operations associated with offshore drilling for oil and gas. In relatively shallow water down to, say, 50m or so, much of this work is done by divers, but the trend today is towards deeper water where diving becomes progressively less economic and increasingly risky. At depths of over 100m the balance of advantage certainly lies with complete automation. In depths measured in hundreds of metres, beyond the continental shelf, such as the Sigsbee Deep in the Gulf of Mexico, where drilling for oil will certainly be taking place by the end of the century, remote control by acoustic means will provide the only possible method. One of the companies, among a growing number, which is actively concerned in development of this kind is Raytheon; it has been working on acoustic systems ever since the very early days at the beginning of the century and has gained much experience from a close association with the US Navy. Its latest concept, known as RATAC (Raytheon Acoustic Telemetry and Control) is a versatile system designed for primary back-up remote control of blowout preventer (BOP) stacks, and can provide monitoring and control functions for multiple wellheads. For the latter, the system can transmit quantitative information from temperature, pressure and flow-rate sensors and indicate valve positions for ten or more wellheads; it can also pass commands to open or close valves, and to actuate the shear ram on command or after failure to acknowledge repeated alarms indicating malfunction. Obviously, with so much at stake, the command signals must be absolutely secure with no possibility of their being activated by extraneous effects. The system is duplicated and designed to 'fail-safe'.

Acoustic telemetry is thus an essential tool for fishing and fisheries research, for oceanographic and marine geophysical investigations, and for the offshore oil industry.

In France the Institut Français du Petrole (IFP), a research and development organisation covering all aspects of the petroleum industry, has developed an acoustic remote control and telemetering system known as TELTAC which has functions broadly similar to RATAC. It is a frequency-modulated system with a designed range of 8km, and can accommodate a large number of coded signals and commands. Other telemeters for naval and oceanographic use have been produced by Thompson CSF.

In Japan the emphasis has been on fishing trawl telemetry. Furuno is one of three Japanese firms manufacturing essentially similar equipment. Frequency modulation of a carrier of 55kHz is used and a range of 2,000m has been fully established in a successful commercial production, certainly the world's largest. The information transmitted is displayed on an echosounder type of recorder and shows everything below the headline in analogue form, ie, fish, ground-rope and seabed. A second channel indicates seawater temperature.

COMMUNICATION AND CONTROL

Future development is likely to concentrate on better methods of locking on to the incoming signal, especially when the relative bearing is changing rapidly and in the presence of severe multipath interference effects; this applies particularly to the needs of the fishing industry. Efforts are being made also to improve the information rate, which, in the present state of the art amounts, in digital language, to a few hundred bits per second. However, to keep up with the greater information rate made possible by high-speed sector scanning techniques, something much higher than this will be required in the future. To transmit the information from a digital sector-scanning sonar mounted on the trawl, something of the order of 10^4 bits/sec will be required. The development of a more sophisticated telemetry system to achieve this is now being undertaken by the University of Birmingham, with the support of MAFF. Even this data rate is insufficient for some purposes and further ideas, such as data compression techniques, are being considered.

THE MOHOLE PROJECT

In March 1961 the first phase of an attempt to drill right through the earth's crust to the mantle was undertaken off Guadeloupe in an area where the water is about 2 miles deep. The technical difficulty of drilling in so great a depth is compensated by the fact that the earth's crust is much thinner under the oceans than it is on land—about 4 miles instead of 25 to 45. The operation was called the Mohole project after the geologists' name for the upper crust of the mantle, the Mohorovicic discontinuity. In the first attempt, made with a drillship named *Cuss I*, samples of basalt rock were brought up from 1,000ft below the seabed. The estimated costs of the further stages were so high that the project was abandoned. Looking back on it, it is probably true to say that the whole concept was rather too ambitious for the technical means available at the time.

This imaginative piece of US Government research, though

unsuccessful in itself, had valuable side effects in providing the money for industry to develop the techniques for drilling in very deep water. One of the most important of these techniques is to hold the drillship or platform rigidly in position over the hole while drilling operations are in progress. In depths of water up to 100m or so this is best accomplished by semi-submersible or jack-up rigs which rest on the bottom when on site. At greater depths floating rigs can be held in position by six or more heavy moorings, the position being monitored and adjusted as necessary with the aid of a taut wire and inclinometer. In still deeper water the inclinometer method becomes unreliable, and eventually the moorings have to be replaced by lateral thrust propellers for station-keeping. The only method of control is an acoustic position referencing system, for which, in principle, there is no ultimate depth limitation.

For the Mohole project, Honeywell of Seattle had the contract, under Brown & Root, for the development of a Dynamic Ship Positioning System. They studied a wide variety of possible methods and ultimately delivered two deep-water acoustic position reference systems for Mohole. One involved the setting up of a group of transponders on the seabed to provide a long base line. The time intervals between the arrival at the ship of the responder transmissions could be computed to fix the position relative to the transponders.

The other system, which was the one subsequently developed by Honeywell for commercial applications, uses the same geometry turned on its head. A single beacon is placed on the sea floor and the time intervals are achieved by recording the signal in two pairs of hydrophones mounted in the fore and aft and athwartships planes. One hydrophone being common to both axes, only three are needed, but a fourth is normally fitted as a stand-by. For a drillship the length of the athwartships base line can be increased by supporting the outside hydrophone on the end of a boom, but for an oil rig there is ample distance in both directions. In this arrangement the beacon transmits a continuous signal, permitting the use of a narrow-bandwidth receiver to cut down background, an important consideration in a potentially noisy environment. For the same reason the hydrophones would normally be mounted as far below the hull as practicable. Differences in the slant ranges from the beacon to the hydrophones are measured as phase differences, and the results are computed by phase-comparison methods similar to those used in radio navigation position-fixing systems. With a baseline of 40ft an accuracy of 1 per cent of the depth of water is claimed by the manufacturer, and beacons with a life of 200 days and capable of withstanding water pressure at depths up to 3,000ft are available.

The ship's position is continuously displayed relative to the beacon, or to an offset position directly over the wellhead on a crt. In a further development the information can be put through a computer to control the engines, or winches, directly to keep the ship automatically locked in its position. In this case, of course, a great deal of data relevant to the particular ship or rig must be fed into the computer. The effects of wind, weather and tide will also have to be allowed for.

ACOUSTIC POSITIONING AND RE-ENTRY FOR DRILLING

The Honeywell system is used in the *Glomar Challenger*. This outstandingly successful deep sea drilling project has now circumnavigated the globe and obtained a series of priceless cores of the sediment and underlying rock in all the deep ocean basins. Penetrations approaching 5,000ft (1·5km) in depths of water up to 20,000ft (6km) have been achieved. The Scripps Institution of Oceanography is managing the project under a contract with the US National Science Foundation. The original programme was extended until 30 June 1973.

During the earlier legs of the programme it was discovered that a layer of chert, a very hard form of limestone, is present over wide areas of the Atlantic and Pacific oceans at a certain depth within the sediment. The drilling bit is usually unable to penetrate this layer without replacement and, in the absence of a relocation sonar designed for borehole re-entry at great depths, some sedimentary profiles had to be left uncompleted. This deficiency was remedied before the north Atlantic leg in 1970.

The re-entry sonar designed by the EDO Western Corporation works in the following way. An up-ended cone, 14ft (4·2m) high and 16ft (4·8m) across its diameter, is lowered on the end of the drill string; it is designed to remain in place over the borehole when the drill pipe is withdrawn to replace the bit. Three acoustic reflectors are mounted round the rim of the cone. For re-entry the drill pipe is lowered to within about 200ft (60m) of the hole and held 30–40ft (9–12m) above the sea floor. The sonar is then lowered down through the drill pipe and scans the seabed to locate the cone reflectors in its searchbeam, which has a maximum range of 500ft (150m). Once the reflectors have been detected on the crt display, the pipe is edged into position over the cone with the aid of the acoustic position reference system, and with additional assistance from a side-thrusting water jet fitted about 60ft (18m) from the bottom of the drill string.

For the final stages of the approach the sonar is switched to the local beam, tilted at a steeper angle and with a wider vertical beam. As soon as the bit has been brought within the area of the funnel, the sonar is quickly retracted and replaced in the pipe by the new bit, so that drilling can be continued in the same hole.

Looking to the future, it is evident that the main demand for this type of equipment will come from the oil industry. The trend in the last few years has been towards offshore drilling in deeper and deeper water; undoubtedly this will continue as fields are discovered and exploited further from land and, within the next 20 years or so, beyond the limits of the continental shelf areas. At present almost all offshore drilling is taking place in relatively shallow depths where the balance of advantage still lies with mechanical rather than acoustic control. So the demand for dynamic positioning, re-entry systems and acoustic BOP control is developing rather slowly.

The pattern of things to come can be seen in the design of the drillship *SEDCO 445*, commissioned in October 1971. Emphasising the international flavour of this industry, the vessel (length oa 445ft) has been designed and built to Shell International specification by the South Eastern Drilling Company (SEDCO) of Dallas, Texas, in the Mitsui yard at Tamano in Japan at a cost of nearly £7 million. Her first assignment is on charter to Brunei Shell, for which company she is carrying out drilling operations in progressively deeper water to test the complete range of remote control acoustic equipment of the kinds just described. The immediate objective is said to be a depth of 2,000ft (600m). Others will certainly follow the lead.

First the Mohole project and then the voyage of *Glomar Challenger* have given the US industry a head start in this important and potentially profitable branch of acoustic technology, but it is not quite alone in the field. In France IFP has sponsored the development, in conjunction with Thompson CSF, of a dynamic positioning system, using acoustic methods. This is the reverse of the one favoured by Honeywell, ie, four transponder beacons (Type TSM 7110) are placed on the seabed in two pairs mutually at right-angles. This arrangement has the advantage over the Honeywell system of providing a longer base line and of having to fit only one hydrophone below the drillship or platform. On the other hand, the distances between the transponders are not known with the same precision as those between the hydrophones in the Honeywell arrangement. However, both manufacturers claim the same order of accuracy—1 per cent of the depth.

The drilling unit of IFP has also developed a re-entry system that may be used in depths of water of over 8,000ft (2·4km). This is based on a sonar system that makes use, as far as possible, of existing equipment. The ship's acoustic dynamic position-fixing equipment is used in the re-entry procedure to locate the ship over the drill hole. The drill string is then lowered and the sonar switched on to detect both the drill string and the wellhole marked by a 20ft funnel. The differences between the X and Y co-ordinates of the two echoes on the ppi scope are then calculated and fed to the dynamic positioning equipment, which automatically moves the ship until the two coincide exactly. Shallow water trials off Marseilles in 1970 confirmed that the drill string could be manoeuvred into position in this way without the use of a jet stream. The sonar was able to detect a 6in diameter drill pipe at 40m. The system is adaptable to other offshore drilling manoeuvres, such as guiding the flexible hose of a tanker to the opening of a submarine storage tank on the seabed.

In 1973, as this book goes to press, GEC-Elliott Electrical Products Ltd UK announced the design of a computer-based dynamic position control system for the new drilling vessel *Wimpey Sealab*. It is required to hold the ship within a circle of 7m radius or 3 per cent of water depth in wind speeds up to 25 knots, currents up to three knots, wave heights of 3·5m and wave lengths of 91m. At wind speeds of 40 knots the limiting radius is extended to 11m. The acoustic reference data is obtained from signals transmitted by beacons on the sea floor to a pattern of hydrophones on the underside of the hull of the vessel. The system will be operational in 1974.

VOICE COMMUNICATION

When a diver is attached by an air or life line, it is evidently most convenient to incorporate in it a telephone wire for communication, and this is done for example, in the Siebe Gorman/McMurdo 'Duck Set'. However, for Scuba (Self contained underwater breathing apparatus) divers a wireless link is more appropriate. Speech can be transmitted over short distances by direct transmission via an amplifier; this is directly comparable to a public address system in air. Alternatively the voice can be modulated with some carrier, ultrasonic sound being the most effective. Frequencies between 8kHz and 150kHz have been used for this purpose and ranges from a few hundred metres to several kilometres have been achieved.

But for the man underwater the essential problem of communication does not lie with the link in the water; it starts with his speech and ends with his listening to, as opposed to merely hearing, what is being said. The diver is always in a very hostile environment, and is subject to several major physiological handicaps, such as cold, anxiety, narcosis, carbon dioxide build-up and anoxia. These difficulties slow down and distort the mental processes and have a cumulative effect that can sometimes be very serious. But this is not all. The process of forming words and speaking clearly is more difficult underwater for several reasons. Chief among these is the limited volume of air in the face mask: when completely submerged, there is negligible transmission through the walls, and this enclosed cavity produces a high acoustic impedance at the lower frequencies used in speech, with a resulting loss of intelligibility. The effect can be mitigated by increasing the volume of the face mask, but this leads to the danger of a build-up of carbon dioxide. There are two possible solutions. One is to introduce a thin, acoustically transparent inner lining to cover the mouth and nose, and the other is to introduce an acoustic feedback path in the form of an amplifier and loudspeaker inside the mask.

Other, less severe, impediments to clear speech are caused by the difficulty of using the facial muscles properly under pressure and against the mechanical restriction of the mask. Finally, in deep water there is the well known 'Donald Duck' effect, caused by the substitution of helium for nitrogen in the air mixture. The speed of sound in helium is several times faster than it is in nitrogen so that the resonant frequencies of the cavities in the head and larynx are correspondingly raised to a squeaky pitch. To bring the speech resonances back to normal without slowing up the words involves a complex process of unscrambling.

This can now be done electronically, and expensively, in real time to give a reasonably intelligible but far from perfect replica. A time domain speech processor recently developed at the Admiralty Research Laboratory has been successfully used throughout a simulated dive to 1,500ft (450m). The Standard Telecommunication Laboratories Ltd, at Harlow, has also designed a processor using the 'vocoder' or an analysis/synthesis technique.

BIO-ACOUSTICS

In striking contrast to the difficulties man experiences in trying to communicate in the hostile environment of the sea, the marine mammals have achieved what seems to be a perfect communion with their relatively new environment; for, in evolutionary terms, it is only recently that these species have returned to the sea and begun to evolve, among other attributes, an acoustic communication system, the refinements of which are, even today, far from fully understood. Indeed, the science of underwater acoustics probably still has much to learn from these fascinating creatures. The ramification of the bone structure and associated membranes that have to do with sound production and reception are extremely complex, and certainly contain as yet undiscovered subtleties.

Among the several species of marine mammal, the *Odontocetes* have the most highly developed sound system. These are the toothed mammals, including the sperm and killer whales, porpoises and dolphins. The sperm whale is acoustically the most advanced of all, but because it is too large to study in captivity, much of the knowledge of its performance comes by inference from an examination of its immense misshapen head, which accounts for over a third of its total length. The forehead is believed to be crucial for echolocation. Each inner ear is encased in dense ivory-like bone delicately suspended by ligaments inside a cavity filled with a sort of stable mucus, something like foam rubber, which provides ideal acoustic insulation. These mammals, in common with other species, make two kinds of sound. One consists of series of millisecond pulses of 'white' noise, peaking at 20–30kHz, used for locating food and obstructions, and the other of pure tone sounds of longer duration, used for communication. There is some evidence to show that they are capable of producing both kinds of sound simultaneously, one from each side of the head.

It has been known for many, perhaps hundreds of, years that whales and other sea mammals can make sound of many kinds, and the annals of the whaling industry are punctuated with picturesque descriptions of them. Whales were heard to 'sing', and there is the well known description, dating from the last century, of 'Kelly's band'. All these accounts relate, of course, to sounds heard in air. The study of underwater sounds dates from the development of sonar, and has only been started in earnest during the last 25 years.

McBride, the first curator of the Marine Studios Oceanarium at St Augustine, Florida, noticed in 1947 that bottlenose porpoises would shy away from an invisible fine mesh net, and postulated echolocation. Since then the study of marine bio-acoustics has expanded rapidly, especially in America. At the time of the second symposium on Marine Bio-acoustics held in New York in 1966 there were already 170 basic research projects in progress, involving an expenditure counted in millions of dollars per annum.

The study of the sounds made by the marine mammals and of the way in which they are used poses a number of environmental problems. Work under controlled conditions must be carried out with the smaller members of the species such as dolphins and porpoises in tanks and aquariums. However, whereas the open sea is essentially a noisy anechoic place, an aquarium is exactly the opposite. It has been noticed that some of the characteristic sounds made by dolphins are absent from their vocabulary in a pool and those used are generally of shorter duration.

The speed of sound in water being about four times as high as it is in air, the wavelength at any particular frequency is correspondingly elongated, and it is therefore necessary to go to higher frequencies in water to obtain comparable discrimination. This is evidently why the auditory apparatus of these mammals is not limited to the sonic region but extends far beyond it. The ambient noise levels in the sea are also much reduced in the ultrasonic regions. Porpoises have been trained to come for a reward to pure sounds up to 150kHz, although the response becomes ragged after about 120kHz. According to workers in the Black Sea, the Afalina dolphin has an upper threshold of 180kHz and can distinguish between glass and foam rubber objects by ultrasonic location methods. Such sounds, of course, can only be examined at secondhand through the medium of a crt. For classifying and differentiating between complex sounds, special instruments, such as Sperry's SPECTRON (Spectral Comparative pattern analyser), are used. Even so, it is possible that the briefest transients are missed, so that the sounds, when played back, lose their significance and meet with a negative response.

A large number of such sounds have been identified and some tentative associations made. Sharply falling, fall-rise-fall and lumped frequency contours appear to be associated with alarm. Some writers with an instinct for dramatic scientific discovery have already assumed that dolphins, with their large brains and advanced acoustic equipment, have discovered the gift of speech, and that it is only a matter of cracking the code before we shall be able to join in the conversation. However, a certain scepticism leads me to think that such hopes, like those of a century ago for life on other planets of the solar system, are doomed to disappointment.

Not many years ago a series of blips were observed on a

SOFAR recorder over long periods at such regular intervals that it was assumed the cause must be a fault in the electronic system or a breakthrough of some other man-made sound. However, after all such possibilities had been eliminated, it became evident that these sounds came from some type of marine organism. These blips have since been identified in widely separated areas in both the Atlantic and Pacific oceans. They consist of strongly peaked 20 cycle pulses of a typical duration of 1sec occurring every 10sec in uninterrupted trains lasting several minutes. They also occur in pairs at 22 and 15sec intervals. Some kind of whale was suspected, as the sources of these sounds appeared to move at between 1 and 4 knots, and the seasonal occurrence between between November and March fitted the migrating habits of some species. Eventually, but not before resorting to aircraft searches, the culprit was established beyond doubt to be the fin whale. But the explanation of the blips is still a matter of controversy; they may be a method of echolocation or even, it has been suggested, a way of overloading the acoustic systems of food fish or predators. Certainly they have a high source level of from 1 to 8W. It is not known how they are produced.

Less work has been done on the study of sounds made by fish, though it is now believed possible that all species of fish make noises of some sort. In certain of the more primitive fishing industries, in Malaysia for example, listening for certain species of shoaling fish underwater is a traditional skill that has been passed on from generation to generation. Generally speaking, these sounds are of a low intensity. For some fishes they have been correlated with behaviour patterns, such as competitive feeding, territorial defence, escaping, migration, spawning, chasing, display fighting and courtship.

Snapping shrimps are well known for the row they make—a more or less continuous noise described as crackling, rasping or frying—covering the frequency band from 1 to 20kHz. The intensity is enough to mask any form of sonar within this frequency band, and charts have been produced to show when and where this interference may be encountered. Mussels and barnacles are also capable of low intensity snapping sounds.

Fish sounds can be divided into three groups related to the methods by which they are produced. These are stridulatory, or rubbing, mechanisms; hydrodynamic phenomena such as occur when a school of herrings or sardines suddenly changes direction; and contractions involving the swim bladder, which functions effectively as a highly efficient low-frequency underwater loudspeaker. The grunts, which make a sound like grinding teeth, use the swim bladder as a resonator.

Finally, there is some doubt about the way in which fishes can detect the bearing of a source of sound. To localise a sound source, two pressure receptors are necessary, yet almost all species of fish have only one swim bladder. However, as one scientist has observed: 'It is dangerous to assume that because we cannot find a physical basis for naturally occurring phenomena, that those phenomena cannot be.' The mammals and the fishes of the sea have not studied physics but they manage pretty well with underwater sound, nevertheless—better perhaps than we who have.

NOTES

CHAPTER 1 BASIC CONCEPTS

1. For the purposes of this account it is sufficient to define the concept of a 'match' or 'mis-match' in terms of the acoustic impedances of the two substances (one of them water) that are in contact with each other. Acoustic impedance equals ρc, where ρ (the Greek letter 'r', pronounced 'rho') is the density and c is the speed of sound. These two values are related to each other for any particular substance by the formula $c = \sqrt{\frac{k}{\rho}}$, where k is the volume elasticity. The nearer the acoustic impedances are, the better the 'match' or the greater is the efficiency of the energy transfer across the interface.

2. The speed of sound in water varies from 1,450 to 1,570m/sec. It increases with temperature at a variable rate of about 4·5m/sec/°C, and with salinity at 1·3m/sec per 1 part in 1,000 increase, and with depth due to the effect of pressure at 1·7m/sec/100m. The empirical expression for the speed of sound is:

$$C = 1{,}449 + 4{\cdot}6t - 0{\cdot}055t^2 + 0{\cdot}003t^3 + (1{\cdot}39 - 0{\cdot}012t)(s - 35) + 0{\cdot}017d$$

where s is salinity in parts per 1,000, d is depth below the surface in metres and t is temperature in degrees centigrade.

3. For Hertz equals cycles per sec, named after Heinrich Rudolph Hertz (d 1894), who demonstrated the existence of electromagnetic waves and whose discoveries led to the development of wireless telegraphy.

4. The decibel, named after Alexander Graham Bell (d 1922), who developed the telephone, dates from the early days of transmission line theory. It is a convenient way of expressing the loss per unit length in terms of the ratio of the intensities at each end. It is a logarithmic function, so that successive increments are additive. The dB being ten times the basic ratio, if I_1 and I_2 are two intensities, then

the ratio between them, N, is expressed as 10 $\log_{10} I_1/I_2$dB. If, for example, I_1 is twice the value of I_2, then I_1 is 3dB above I_2 (log 2 = 0·301, so 10 log 2 = 3 approx). The intensity of the sound transmitted into the water by a modern echosounder may well be a million times greater than that of the returning echo—a loss of 60dB between transmission and reception. The directionality of a sound pulse is usually defined in terms of the sector within which the power does not fall below half the value of the peak power in the centre of the beam. This is stated as, say, 15° between 3dB, or half power, points.

5. 'Transducer' is used throughout to mean 'electro-acoustic transducer', that is to say, a device for converting electrical energy into acoustic energy and the reverse. A projector is a device for initiating sound transmission underwater and a hydrophone is a device for receiving sound waves.

6. The extent of the Fresnel zone is conventionally taken as $R^2/2\lambda$, where R is the half-length of the transducer. For most instruments used in underwater detection this only extends for a few feet and is thus too short to have any operational significance.

7. Aristophanes coined a beautiful onomatopoeic word to describe the sound of bubbles rising, and breaking—*πομφολυγοπαφλ-ασμα.*

CHAPTER 2 ANTI-SUBMARINE WARFARE

1. The following different explanations of the word asdic have been published.

Renou & Tchernia. *La Revue Marine Nationale*, No 29 (1947)—Anti-Submarine Device International Committee.

Hillar, A. P. USN. *CNO Publication OPNAV*, P413–104 (1946)—Allied Submarine Devices Investigation Committee.

Lewis, D. D. USN. *The Fight for the Sea* (1961)—Anti-submarine detection indicator.

Wood, A. B. *RNSS Journal* (July 1965)—Anti-submarine Division-ics, as in 'physics' for example.

Gale. *Acronyms and Initials*—Anti-submarine Detection Investigation Committee.

Finally, in reply to a request for an explanation after Winston Churchill had used the word in parliament during World War II, the captain of HMS *Osprey*, the British Naval Anti-submarine school, stated it to be an acronym for the 'Allied Submarine Detection Investigation Committee', formed during World War I, which organised much R & D for submarine detection. No two explanations agree and the records of this committee cannot be traced and probably no longer exist. As to 'sonar', the word coined in World War II by the USN and now generally adopted, there is no doubt about its origin; it stands for SOund NAvigation and Ranging. For the sake of consistency sonar is used throughout this text to denote echoranging systems.

2. British patent 11,125 (1912), filed 10.5.12. Complete specification 10.12.12, accepted 27.3.13.

3. United States patent application 744,793 (1913).

4. Rochelle salt (sodium-potassium tartrate), so named for having been discovered in 1672 by an apothecary living at La Rochelle. It exhibits the greatest known p-e effect, about ten times that of ADP, but has disadvantages as a material. It is fragile, soluble in water and liable to disintegrate when subjected to a high voltage potential.

5. The doppler shift, named after Christian Johann Doppler (d 1853), the Austrian physicist who discovered it, occurs whenever the distance between an observer and a source of constant vibration (light or sound) is changing. The wavelength is increased and therefore the frequency is reduced, whenever the source and the observer are moving away from one another, and vice versa. In the case considered here, however, the observer is aware of the change of pitch between the submarine echo and the background of reverberation, not the pitch of the transmitted signal, which he cannot hear. Reverberation, made up of echoes from targets that are stationary in the water, consequently has a pitch that differs from the pitch of the transmission by the speed of the ship in the direction of the beam. The doppler shift heard by the operator is therefore a function of the speed of the target through the water, not relative to the ship. The actual amount of the shift is proportional to the transmitted frequency. To take a concrete example, a movement of the target of 2 knots along the 'line of sight' would cause a shift of 3·45 per cent for a 25kHz transmission. After heterodyning to an audible frequency of 1kHz this amounts to 34·5Hz, which is about half a semitone and just detectable to a trained ear.

CHAPTER 3 NAVIGATION

1. The relationship between frequency (f) in kHz, total beam angle between 3dB points (θ) in degrees, and wavelength (λ) in cm is defined by the expressions:

$$\lambda = 150/f \text{ and } \theta = k/N,$$

where N is the width of the transducer in wavelengths and k is a constant taken as 51 for a rectangular transducer and 59 for a circular one.

Thus, for a 50kHz transducer the wavelength is 3cm. If it is 1ft (30cm) square, the total beam angle $\theta = 51/10$ or 5·1°.

However, if the frequency is only 12kHz, the wavelength is 12·5cm and, for a transducer of the same size, $\theta = 51 \times 12{\cdot}5/30$, or 21·25°. Conversely, to obtain a 5·1° beam with this frequency, the transducer would have to be 125cm (49in) wide.

2. Beam steering is the technique of shifting the centre bearing of the transducer electronically instead of rotating it physically. To do this the oscillating face is divided into a number of separate elements, each of which is connected to the receiver via a phase-shifting device. By shifting the phase angle progressively across the face of the transducer, the direction of maximum sensitivity, ie, the direction from which a wavefront of an incoming signal is in phase, can be shifted from the centre bearing to any other position, thus simulating the actual movement of the transducer. The principle applies also for the transmission.

3. Such as, Raytheon, EDO Western, Gifft, Alden, Oceansonics, EPC, ORE, Kelvin Hughes, ELAC, among a number of others.

4. The sonar equation strikes a balance between the performance of an echosounder, on the left-hand side, and the operating conditions, on the right-hand side. The former includes the source level, transducer directivity index and the receiver bandwidth, and the latter spreading and attenuation losses in the water and the bottom reflection loss for any given depth. For the purpose of this test the right-hand side of the equation is termed the 'figure of merit' and has been calculated (in dBs), for different frequencies at a depth of 400m.

CHAPTER 4 FISH DETECTION

1. 'It seems probable that dense banks of fish such as sardines and herrings can reflect ultrasonic waves and that one could, by this means, find out the exact depth at which these fish shoals are to be found on the previously sounded continental shelf.'

2. Onazote is foamed ebonite, a substance with an acoustic impedance very similar to that of air.

3. 'A convoy coming from America was crossing the Atlantic on its way to Gibraltar. On each side, in front and behind, the escort vessels were patrolling. On the bridges one heard the monotonous "ding", "ding" of the asdic transmissions. No echoes.

Action alarm. Echo bearing 300°. Range 1,000 yards, signalled the *Sabre* giving chase. . . .
The anxious convoy waits for him to express his opinion. Was it a U-boat pack lying in wait? And, simply, a very calm voice explains, "The target detected was a shoal of fish." '

4. The Natural Environment Research Council Working Group on Underwater Acoustics was appointed on 10 July 1968 with the following terms of reference:

> To promote research and development on acoustic methods for fish detection in the sea.
>
> To identify future needs in this field for both research and commercial purposes, having regard to world markets in fishing and instrumentation.
>
> To recommend to the Oceanography and Fisheries Committee a co-ordinated programme appropriate to meet these needs having regard to the particular interests, expertise and responsibilities of the various organisations concerned.

The membership of the working group consists of representatives of the following organisations:

University of Birmingham (Chairman and two)
White Fish Authority (two)
Department of Agriculture and Fisheries for Scotland (two)
Ministry of Agriculture, Fisheries and Food (two)
Admiralty Underwater Weapons Establishment (two)
National Institute of Oceanography

In addition, part-time members from Bath University of Technology, Unit of Coastal Sedimentation, Marine Technology Support Unit, Department of Trade & Industry and Royal Aircraft Establishment have attended one or more meetings.

BIBLIOGRAPHY

CHAPTER 1

Cady, W. G. *Sound, its uses and control*, 2 (1963), 46–52

Haslett, R. W. G. 'Underwater Acoustics', *J Sc Instrum*, 44 (1967), 709–19

Hunt, F. V. *Electroacoustics; analysis of transduction and its historical background* (Cambridge, Mass, 1964)

Rayleigh, J. W. S. *Theory of Sound* (1877)

Stratton, Julius A. *Our Nation and the Sea*, Report of the Commission on Marine Science, Engineering and Resources (Washington, DC, 1969)

Tucker, D. G. & Gazey, B. K. *Applied Underwater Acoustics* (1966)

Urick, R. J. *Principles of Underwater Sound for Engineers* (New York)

Williams, Jerome. *Sea and Air, the Naval Environment* (US Naval Institute, Annapolis, Md, 1968)

Williams, S. R. 'Magnetostrictive effects', *J Optical Soc Amer* (May 1927), 383

Wood, A. B. *A Textbook of Sound* (1930)

CHAPTER 2

Haslett, R. W. G. *et al.* 'The Underwater Acoustic Camera', *Acustica (17.4.1966)*, 187–203

Hayes, Harvey C. *A brief history of the Sound Division, NRL from 1917 to 1946*, Unpublished ms

——. 'Detection of submarines', *Am Phil Soc* (Jan 1920)

Hilar, A. P. *Sonar—Detector of submerged submarines.* OPNAV P413–104 (Washington, DC, 1946)

Klein, Elias. *Underwater Sound Research and Applications before 1939*, ONR Report ACR–135 (Washington, DC, 1967)

Spitzer, Lyman, *et al. Underwater Acoustic Research*, Committee on Undersea Warfare NRC (Washington, DC, 1950)

Wood, A. B. 'From the Board of Invention and Research to the Royal Naval Scientific Service', *J RN Sci Service*, 20.4 (1965), 200–81

CHAPTER 3

Ahrens, E. 'Methoden der Teefenvermessung mit ultraschall-echoloten', 13th Congress of the International Federation of Surveyors (1971)

Boyle, R. W. & Reid, C. D. 'Detection of Icebergs', *Trans R Soc Canada*, 20 (iii) (1926), 233

Campbell, A. C. 'Geodetic methods applied to Acoustic Positioning', 1st Marine Geodesy Symposium (Washington, DC, 1966)

Cook, J. C. 'Some practical applications and limitations of high definition depth scanning sonars', Norspec Conference (London 1970)

Carr, J. B. 'Doppler Sonar Navigator System for Deep Submersible Vessels', Norspec Conference (London 1970)

Dickson, A. F. 'Underkeel clearance', *Inst of Nav J* (1967)

Marti, M. 'Le Sondage en Mer par le son aux grandes profondeurs au moyen de detonations', *Annales Hydrographiques* (1923–4)

——. 'Note sur le sondage acoustique', *Int Hyd Review*, 8.1 (1931), 133–49

Maul, G. H. & Bishop, J. C., Jr. 'Mean sounding velocity—a brief review', *Int Hyd Review*, 47.2, 85–92

Spiess, F. N. 'Underwater Acoustic Positioning—Applications', 1st Marine Geodesy Symposium (Washington 1966)

Tyrrell, W. A. 'Underwater Acoustic Positioning—Principles and Problems', 1st Marine Geodesy Symposium (Washington 1966)

Walsh, G. M. 'The Finite Amplitude Depth Sounder (FADS)', Eng in the Ocean Environment Conf, *IEEE* (1971)

Wood, A. B. *et al.* 'Magnetostriction Depth Recorder', *JIEE*, 76 (1935), 550–63

Present and Future Civil Uses of Underwater Sound', Committee on Underwater Telecommunications, NRC Nat Ac of Sc (Washington, DC, 1970)

'Recommended Echo Sounder Performance Specification DTI' (1972)

CHAPTER 4

Azhazha, V. G. & Shishkova, E. V. *Fish location by hydroacoustic devices*

BIBLIOGRAPHY

Balls, R. *Fish on the Spotline* (1946)

Budker, Paul. *Whales and Whaling* (1958)

Cushing, D. H. *The uses of echosounders for Fishermen* (1963)

Forbes, S. T. 'A Quantitative fish counting echosounder', IERE Conference Electronic Engineering in Ocean Technology (1970)

Haines, R. G. *Echo Fishing* (1969)

Harden-Jones, F. R. & Pearce, G. 'Acoustic Reflection experiments with Perch to determine the proportion of the echo returned by the swim bladder', *J of Exp Bio*, 35.2 (1958), 437–50

Haslett, R. W. G. 'Determination of the acoustic scatter patterns and cross sections of fish models and ellipsoids', *Br J of Appl Physics*, 13 (1962), 611–20

——. 'Measurement of the dimensions of fish to facilitate calculations of echo strength in acoustic fish detection', *J du Conseil International pour l'Exploration de la Mer*, 27.3 (1962), 261–9

Hodgson, W. C. & Fridriksson, A. 'Report on Echosounding and Asdic for Fishing Purposes', *Rapports du Conseil Permanent International pour l'Exploration de la Mer*, Vol 139 (Copenhagen 1955)

——. 'Echosounding and the pelagic fisheries', *MAFF Fisheries Investigations*, Vol 17.4 (1950)

Hopkins, P. R. 'Cathode Ray Tube displays for Fish Detection on Trawlers', *J Brit IRE*, 25.1 (1963), 73–82

Hopkins, J. C. 'Cathode Ray Tube display and correction of Side-scan sonar signals', IERE Conference on Electronic Engineering in Ocean Technology (1970)

Johnson, H. M. & Proctor, L. W. 'Advanced design of a Fishing Sonar', *Oceanology International* (1972)

McCarthy, W. J. *The Application of Asdics to Whalecatching* ACSIL/ADM/47/173 (1946), Unpublished ms

Mitson, R. B. & Cook, J. C. 'Shipboard installation trials of an electronic Sector Scanning Sonar', IERE Conference on Electronic Engineering in Ocean Technology (1970)

Renou, J. *La Détection des Poissons par les ondes ultrasonores* (Paris 1947)

Sund, O. 'Echosounding in Fishery Research', *Nature*, 135, No 3423 (1935), 953

Tucker, D. G. *Underwater Acoustics*. Report by NERC Working Group. NERC Series C, No 6 (1971)

——. *Underwater Observation using Sonar* (1966)

Tucker, D. G. *et al.* 'Electronic Sector Scanning', *J Brit Inst Radio Eng*, 18 (1958), 465

BIBLIOGRAPHY

Voglis, G. M. & Cook, J. C. 'Underwater applications of an advanced Acoustic Scanning Equipment', *Ultrasonics*, 4 (1966), 1–9

——. 'General Treatment of Modulation Scanning', Part I, *Ultrasonics*, 9.3 (1971); Part II, *Ultrasonics*, 9.4 (1971)

—— 'Design Features of Advanced Scanning Sonars', *Ultrasonics*, 10.1 (1972)

CHAPTER 5

Brunel, M. *Sondage et Repérage Acoustique* (Paris 1943)

Chesterman, W. D. & Wong, H. K. 'Bottom Sediment distribution in certain inshore waters off Hong Kong', *Int Hyd Rev*, 48.2 (1971), 51–69

Chramiac, M. H. & Morton, R. W. 'High Resolution near subbottom Profiling', Offshore Technology Conference, Dallas

Cloet, R. L. 'How Deep is the Sea ?', *J Inst of Nav*, 23.4 (1970), 416–31

Daniels, D. & Henderson, R. 'An integrated acoustic Underwater Survey System', Oceanology Conference (Brighton 1969)

Gauss, G. A. 'Acoustic techniques for Geological studies with particular reference to Dredging problems', Norspec Conference (London 1970)

Glen, N. C. 'Hydrographic Surveying in the North Sea, Norspec Conference (London 1970)

Glenn, M. F. 'Introducing an operational Multi-beam Array Sonar', *Int Hyd Rev*, 47.1 (1970), 35–9

Gray, F. & Owen, T. R. E. 'A Recording Sono-radio Buoy for seismic refraction work', Oceanology Conference (Brighton 1969)

Haslett, R. W. G. & Honnor, D. 'Some recent developments in sideways looking sonars', IERE Conference Electronic Engineering in Oceanography (1966)

Haslett, R. W. G. & Halliday, W. 'Rapid Hydrographic Surveying using Mirror Sonars', IERE Conference Electronic Engineering in Ocean Technology (1970)

Haslett, R. W. G. 'Experiments with a new Mirror Sonar covering a sector of 30°', Oceanology International 72, Brighton

Hatfield, H. R. *Admiralty Manual of Hydrographic Surveying*, Vol 4 (1968)

Heaton, M. J. P. & Haslett, R. W. G. 'Interpretation of Lloyd Mirror in sidescan sonar', Seminar, Bath Univ of Tech (1971)

Hutchins, R. W. 'Broadband Hydro-acoustic sources for High Resolution Sub Bottom Profiling', Oceanology 69 Conference, Brighton

Kawakami, K. & Sato, K. 'Hydrographic Survey for Deep Draught Vessels', 13th Congress Int Fed of Surveyors (Wiesbaden 1971)

Krause, V. F. & Kanaev, V. F. 'Narrow beam Echosounding in Marine Geomorphology', *Int Hyd Rev*, 47.1 (1970), 23–32

Leenhardt, O. 'Mud Probing', *Int Hyd Rev*, 46.2 (1969)

Roberts, W. J. M. 'Large Scale Offshore Surveying for the Oil Industry', *Int Hyd Rev*, 47.2 (1970), 41–64

Roberts, W. J. M. & Weeks, C. G. McQ. 'Hydrographic Automation—A Progress Report', 13th Congress Int Fed of Surveyors (Wiesbaden 1971)

Roberts, W. J. M. 'Charting a telephone cable', *Hydrospace*, 4.2 (1971), 47–50

Rusby, S. 'A Long Range Sidescan Sonar for use in the deep sea (GLORIA Project)', *Int Hyd Rev*, 47.2 (1970), 25–39

Sargent, G. E. G. 'Application of Acoustics and Ultrasonics to Marine Geology', *Ultrasonics*, 6.1 (1968), 23–8

Savit, Carl H. 'Multiple reflection in Marine Seismic Prospecting', 1st Int Congress Petroleum and the Sea (Monaco 1965)

Slattery, F. L. 'A new Hydrographic Data Acquisition System HYDAS)', 13th Congress Int Fed of Surveyors (Wiesbaden 1971)

Sothcott, J. E. L. & Benning, N. H. 'Heave correction in Echosounding', IERE Conference on Electronic Engineering in Ocean Technology (1970)

Stride, A. H. 'Current-swept sea floors near the Southern half of Great Britain', *J Geological Soc of London*, 119 (1963), 175–99

Stubbs, A. R. 'Identification of patterns on Asdic Records', *Int Hyd Rev*, 40.2 (1963), 50–68

Ulonska, A. 'Genauere Seevermessung durch direkte Messung der Schallgeschwindigkeit im Wasser', 13th Congress Int Fed of Surveyors (Wiesbaden 1971)

Voglis, G. M. & Cook, J. C. 'A new source of Acoustic Noise observed in the North Sea', *Ultrasonics*, 8.2 (1970), 100–1

Weeks, C. G. McQ. 'The use of a dual frequency Echosounder in sounding an irregular bottom', *Int Hyd Rev*, 48.2 (1971), 43–9

Winstanley, J. D. 'Surveying Deep Draught Routes', *Inst of Nav J*, 23.4 (1970), 411–16

White, J. C. E. 'Hydrographic and Tidal Information for Deep Draught Ships', 13th Congress Int Fed of Surveyors (Wiesbaden 1971)

Wentzell, H. 'Bodenkartenschreiber', 13th Congress Int Fed of Surveyors (Wiesbaden 1971)

Wong, H. K. & Chesterman, W. D. 'Comparative Sidescan Sonar and Photographic Survey of a coral bank', *Int Hyd Rev*, 47.2 (1970), 11–23

CHAPTER 6

Berktay, H. O. & Gazey, B. K. 'Communication aspects of Underwater Telemetry', IERE Conference Electronic Engineering in Oceanography (1966)

Brooke, J. P. & Mason, C. S. 'Some instruments for monitoring the performance of Undersea mechanical devices', Oceanology Conference (Brighton 1969)

Cattanach, D. 'A 16-channel Digital Acoustic Telemetry system', IERE Conference Electronic Engineering in Ocean Technology (1970)

Harris, M. J. 'Acoustic Command System', Oceanology Conference (Brighton 1969)

Hearn, P. J. & Berktay, H. O. 'Underwater Telemetry and Communications', Oceanology Conference (Brighton 1969)

—— ——. 'Underwater Communication, Past, Present and Future', *J of Sound and Vibration*, 7.1 (1968), 62–70

Heinmiller, R. H. *Acoustic Release Systems*, Woods Hole Oceanographic Institution, Ref 68–48, Unpublished ms

Norris, K. S. *Whales, Dolphins and Porpoises* (New York)

Ray, R. 'Communications between Divers', Oceanology Conference (Brighton 1969)

Tavolga, W. N. *Marine Bio-Acoustics* (New York 1964)

ACKNOWLEDGEMENTS

TO WRITE A BOOK OF THIS NATURE IS BOTH A HUMBLING AND rewarding experience. Humbling because the limits of one's knowledge are pitilessly exposed, and rewarding for the discovery of the extent to which both personal friends and professional librarians are willing to take time and trouble to correct, assist and advise. My thanks are especially due to Bill Halliday, who has been through the whole draft with his critical eye, and whose astringent and felicitous comments have saved me from many errors and imbecilities; also to Jock Anderson, Deryck Chesterman and John Roberts, who have read different chapters and have given me much valuable advice and information.

Of the librarians and staffs who have been unfailingly helpful and patient, I should like to mention in particular those of the Ministry of Defence Library at Lillie Road in the Empress State Building, the Hydrographic Library at Taunton, the Defence Research Information Centre and the Service Hydrographique et Oceanographique de la Marine in Paris.

Finally, there are many other old colleagues and friends who have met my every request and who will, I hope, forgive me for not mentioning them all individually; the list would be too long.

INDEX

Italic numerals refer to illustration pages

INDEX